SMP interact

C3

Book C3

GILBERD SCHOOL
Brinkley Lane, Colchester

CAMBRIDGE
UNIVERSITY PRESS

PUBLISHED BY THE PRESS SYNDICATE OF THE UNIVERSITY OF CAMBRIDGE
The Pitt Building, Trumpington Street, Cambridge, United Kingdom

CAMBRIDGE UNIVERSITY PRESS
The Edinburgh Building, Cambridge CB2 2RU, UK
40 West 20th Street, New York, NY 10011-4211, USA
477 Williamstown Road, Port Melbourne, VIC 3207, Australia
Ruiz de Alarcón 13, 28014 Madrid, Spain
Dock House, The Waterfront, Cape Town 8001, South Africa

http://www.cambridge.org

Printed in the United Kingdom at the University Press, Cambridge

Typeface Minion   *System* QuarkXPress®

A catalogue record for this book is available from the British Library.

ISBN 0 521 78536 7 paperback

Typesetting and technical illustrations by The School Mathematics Project
Illustrations on page 5 (A6) and page 8 by Simon Hayes
and on page 5 (A11) and page 9 by Valerie Grace
Other illustrations by Robert Calow and Steve Lach at Eikon Illustration
Cover image © ImageState Ltd
Cover design by Angela Ashton

The publishers would like to thank the following for supplying photographs:
Page 5 English Heritage Photographic Library.
Page 122 Graham Portlock

The authors and publishers would like to thank Hywel Sedgwick-Jell for his help with the
production of this book.

Contents

① Circumference of a circle

This work will help you

♦ calculate a circle's circumference from its diameter or radius

♦ calculate its diameter or radius from its circumference

♦ solve problems that involve these measurements

A How many times?

The **circumference** of a circle is the distance all round it.

• Explain why diagram A shows that the circumference is more than 3 times the diameter.

• Explain why diagram B shows that the circumference is less than 4 times the diameter.

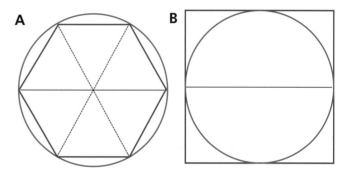

• Would you say that the circumference is closer to $3 \times diameter$ or $4 \times diameter$?

A1 Roughly how much sticky tape is needed to go once round the curved part of each of these parcels?

(a)

15 cm

(b)

7.5 cm

(c)

23 cm

A2 A wedding ring has a diameter of 1.9 cm.

Roughly how long would the ring be if it was cut and straightened out?

A3 A food manufacturer wants to design a label for this size tin.

(a) Roughly how long does it have to be?

(b) How wide does it have to be?

9.5 cm

10.8 cm

A4 A reel of cotton has 1000 turns on it.
The reel has a diameter of 3 cm.

Roughly how much cotton is on the reel?

A5 This is a design for an earring made from silver wire.
About how much wire will be needed?

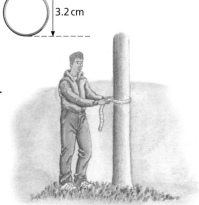

A6 The trunk of a tree has a roughly circular cross-section.
The distance all round the trunk is 135 cm.

What is the diameter of the trunk, roughly?

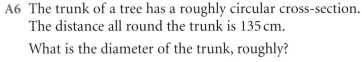

A7 At Avebury, in Wiltshire, there is a prehistoric circle of stones.
The distance all round the circle is 1150 m.

What is the diameter of the circle, roughly?

A8 The distance round the equator is about 25 000 miles.
What is the diameter of the Earth, roughly?

A9 A piece of wire 6 metres long is bent into
the shape of a circle.
What is the **radius** of the circle, roughly?

A10 Silbury Hill in Wiltshire is a prehistoric mound with a circular base.
The distance round the bottom of the hill is 530 m.

In 1968, archaeologists dug a tunnel at the base,
to the centre of the hill.
Roughly how long was their tunnel?

A11 10 people stand in a circle holding hands like this.

  (a) What will its diameter be, roughly?

  (b) Without finding the circumference,
     how can you find the diameter for 20 people?

  (c) What is the diameter for 30 people?

B Becoming more accurate

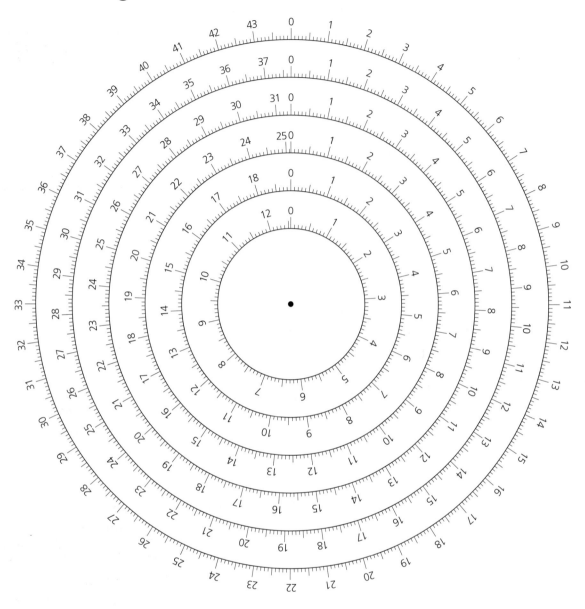

B1 The circumferences of these circles are marked off in centimetres and tenths of a centimetre.

Copy this table and fill in the circumference and diameter for all six circles.

For each circle, find a number that you multiply the diameter by to get the circumference.

Write each 'multiplier' in the table.

Diameter	×?	Circumference

Your 'multipliers' in B1 should be between 3.13 and 3.15.

The exact number to multiply by is called π (pronounced 'pie').

If you press the π key on your calculator, you will get a value with many decimal places, far more accurate than you need for everyday purposes.

$$3.141592654$$

Formulas for circumference

Let r be the radius of a circle,

d the diameter,

C the circumference.

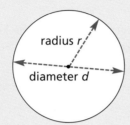

Then $C = \pi d$

Also, because $d = 2r$, it follows that

$$C = \pi \times 2r$$

which is written

$$C = 2\pi r$$

Use the π key on your calculator for the following questions.

B2 Calculate, to the nearest 0.1 cm, the circumference of a circle whose diameter is

(a) 4.4 cm   (b) 3.4 cm   (c) 2.8 cm   (d) 1.8 cm   (e) 0.9 cm

B3 The radius of this circle is 1.3 cm.

(a) Write down the diameter.

(b) Calculate the circumference, to the nearest 0.1 cm.

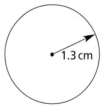

B4 Calculate, to the nearest 0.1 cm, the circumference of a circle whose radius is

(a) 1.9 cm   (b) 1.4 cm   (c) 3.6 cm   (d) 0.8 cm   (e) 5.9 cm

B5 Calculate, to the nearest 0.1 cm, the circumference of a circle with

(a) diameter 6.4 cm   (b) radius 12.8 cm   (c) diameter 10.5 cm

(d) radius 23.2 cm   (e) diameter 15.2 cm   (f) radius 15.2 cm

B6 For each of these circles, measure and record the diameter and then work out the circumference to the nearest 0.1 cm.

(a) 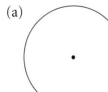   (b) 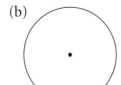   (c)   (d)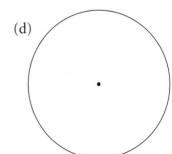

B7 (a) The minute hand on a town hall clock is 1.80 m long.
How far does its tip travel in 1 hour?
(Give your answer to the nearest cm.)

(b) The hour hand is 1.20 m long.
How far does its tip travel in 1 hour?

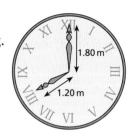

B8 A circular cycle track has an inside radius of 100 m.
The track is 8 m wide.

How much further does a cyclist go in one lap if she keeps
to the outside of the track rather than the inside?
(Give your answer to the nearest metre.)

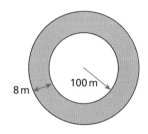

Towards greater accuracy

A round number

The Bible describes a 'sea of cast metal' in
Solomon's temple, with these dimensions.

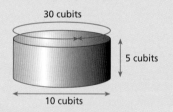

(A cubit is about half a metre.)

This suggests π is 3, the rough value you
used earlier.

Egypt and Babylon

Nearly 4000 years ago the Egyptians
used a value of 3.16.
The ancient Babylonians used 3.125.
You can get this degree of accuracy from
measuring, as in sections B and D.

π from polygons

For hundreds of years people found
more accurate values of π by
calculating the perimeter of polygons
inside and outside a circle, doubling
the number of sides each time.

In 1593 a Dutchman, Adriaen Romanus, found π to
15 decimal places, by considering a polygon with
over 100 million sides!

Billions of decimal places

With the help of computers, π has
been calculated to many billions of
decimal places.

Mathematicians have proved that,
however many decimal places you
calculate, you will never get the
exact value of π.

3.14159265358979323846264338
3279502884197169399375105820
9749445923078164062862089986
2803482534211706798214808651
3282306647093844609550582231
7253594081284811174502841027
0193852110555964462294895493
0381964428810975665933446128
47564823

C Calculating diameter and radius

The formula $C = \pi d$ can be shown as a flow diagram.

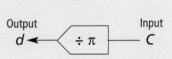

Reversing the diagram leads to a formula for finding d.

$$d = \frac{C}{\pi}$$

The formula $C = 2\pi r$ can be shown as a flow diagram.

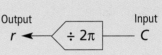

Reversing the diagram leads to a formula for finding r.

$$r = \frac{C}{2\pi}$$

C1 Calculate, to the nearest 0.1 cm, the diameters of circles that have these circumferences.

(a) 4.9 cm   (b) 7.2 cm   (c) 8.5 cm   (d) 9.8 cm

C2 Sean wanted to calculate the radius of a circle of circumference 58 cm.
Here is the key sequence he used on his calculator.   5 8 ÷ 2 × π =
Is this sequence correct? If not, what should it be?

C3 Calculate, to the nearest 0.1 cm, the radius of a circle whose circumference is

(a) 67 cm   (b) 146 cm   (c) 44.8 cm   (d) 106.2 cm   (e) 9.4 cm

C4 Petra has a piece of silver wire 20.0 cm long.
She wants to make it into a circular bracelet. What diameter will it have?

C5 These are descriptions of three circular plates.

Plate P has radius 12.4 cm.   Plate Q has diameter 20.4 cm.

Plate R has circumference 66.6 cm.

(a) Which plate is biggest?   (b) Which is smallest?

C6 An iron strip 5.30 metres long is made into a circular band
to go round a wagon wheel.
What is the diameter of the wheel?

D Rolling and turning

Discuss this experiment
and take any measurements
you think you will need to
predict exactly what will happen.

Then carry out the experiment to
see if you were right.

A large sheet of scrap paper,
resting on a board, is raised up
to a slight slope.

A small mark is made with lipstick on a drinks can,
which is placed at the top of
the slope and allowed to roll down.

D1 The wheels of a tricycle have diameter 40 cm, including the tyres.

(a) What is their circumference?

(b) How far forward, in metres, does a wheel go in 100 turns?

(c) What is the tricycle's speed in metres per second if
the wheels are turning at 100 turns per minute?

D2 An odometer or click wheel is used by surveyors
to measure short distances.

The wheel goes round once for every metre the wheel is pushed.
Calculate its radius, to the nearest 0.1 cm.

D3 The wheels of a car have a diameter of 48 cm.

(a) How far does the car move forward as the wheels make
one complete revolution?
Give the answer in metres to two decimal places.

(b) How many turns do the wheels make when
the car goes 1000 metres?

D4 A burglar on a getaway bike rides over a blob of paint.
It leaves a pattern of blobs like this.

← 207 cm →

What is the diameter of the bike's wheels?

D5 The winch drum at the top of a well has a radius of 13 cm.
The water in the well is 9 metres down.

How many turns of the handle are needed to bring up a bucket of water?

*D6 A drill bit with diameter 6 mm is turning at 900 revolutions per minute.
Find the speed of a point on the edge of the bit in metres per second.

E Parts of circles

E1 12 pieces of model railway track like this make a circle.

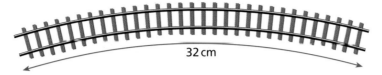

32 cm

Calculate the radius of the circle.

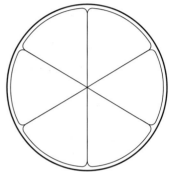

E2 The perimeter of a shape is the distance round its outside.
This carton of cheeses has a radius of 5 cm.

What is the perimeter of one piece of cheese?

E3 This shape is made from three-quarters of a circle.
Find its perimeter.

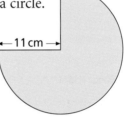

←—11 cm—→

E4 This shape is made from three half-circles
with their centres on the dotted line.

What is the shape's perimeter?

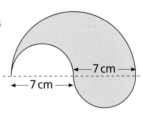

—7 cm—

←—7 cm—→

*E5 This shape is a semicircle (a half-circle).
Its perimeter is 12.3 cm.

What is its diameter?

Explain carefully how you got your answer.

*E6 These three circles are all different sizes.
Their centres are all on the dotted line.

What is the total of all their circumferences?

Explain carefully how you got your answer.

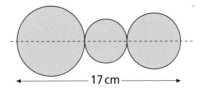

←————— 17 cm —————→

What progress have you made?

Statement

Evidence

I know what the words 'diameter' and 'radius' mean.

1 Measure the diameter and radius of this circle.

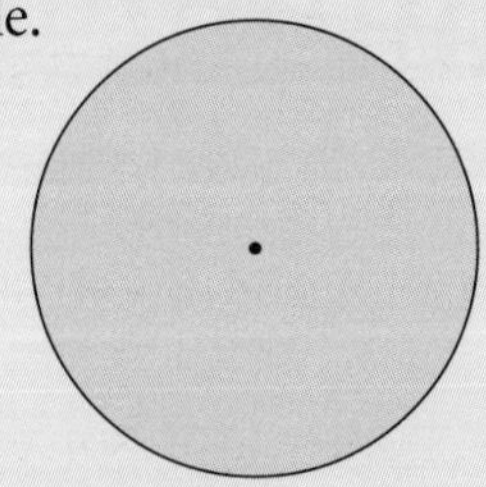

I can calculate a circumference from a diameter or a radius.

2 What is the circumference of a circle with diameter 2000 metres?

3 What is the circumference of a circle with radius 2.8 cm?

I can calculate a radius or a diameter from a circumference.

4 What is the diameter of a pole with circumference 37 cm?

5 What is the radius of a circular pond with a path round it 100 metres long?

I can solve circumference problems that involve rolling or turning.

6 A bicycle wheel's diameter is 72 cm. How far does the wheel move forward during one revolution?

7 The rocket cars on a fairground ride go round in a circle.
They are 7 m from the centre of the ride.
At top speed the ride makes one complete turn every 10 seconds.
What is the top speed of a rocket car in metres per second?

I can solve problems involving part of a circumference.

8 This shape is a quarter-circle. Find its perimeter.

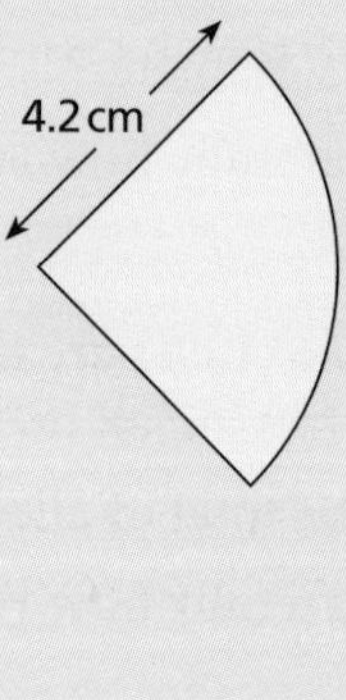

4.2 cm

② Solving equations

This work will help you

◆ form and solve linear equations

◆ solve simple non-linear equations

A Balancing

There is more than one way to solve $2(x + 3) = 10$.

Julie Dodd

$2(x + 3) = 10$

$2x + 6 = 10$ (multiplying out brackets)

$2x = 4$ (– 6 both sides)

$x = 2$ (÷ 2 both sides)

Donal Ranger

$2(x + 3) = 10$

$x + 3 = 5$ (÷ 2 both sides)

$x = 2$ (– 3 both sides)

- Whose method do you prefer?
- How would you solve $3x + 2 = x – 6$?

Julie Dodd

$3x + 2 = x – 6$

$3x + 8 = x$ (+ 6 both sides)

$2x + 8 = 0$ (– x both sides)

$2x = {}^-8$ (– 8 both sides)

$x = {}^-4$ (÷ 2 both sides)

Donal Ranger

$3x + 2 = x – 6$

$2x + 2 = {}^-6$ (– x both sides)

$2x = {}^-8$ (– 2 both sides)

$x = {}^-4$ (÷ 2 both sides)

A1 Solve the following equations.

(a) $7x – 5 = 30$ (b) $9x + 5 = 29 + x$

(c) $3x + 2 = 5x – 1$ (d) $7x – 28 = 9x – 20$

A2 This is how Donal solved the equation $3(x – 1) = 105$.

Describe what he did to both sides of the equation to solve it.

$3(x – 1) = 105$

$x – 1 = 35$

$x = 36$

A3 Julie and Donal start to solve $4(x + 3) = 28$.

 I'm going to multiply out the brackets to start with. I'm going to divide both sides by 4.

(a) Solve the equation $4(x + 3) = 28$ using

  (i) Julie's method   (ii) Donal's method

(b) Which method do you prefer? Which is quickest?

A4 (a) Solve the equation $6(x - \frac{1}{2}) = 3$ in two different ways.

  (i) Multiply out the brackets to start with and then solve the equation.

  (ii) Divide both sides by 6 and then solve the equation.

(b) Which method do you prefer?

A5 Solve these equations.

(a) $5(x - 3) = 65$ (b) $6(w + 5) = 39$ (c) $4(7 + y) = 16$

(d) $8(2z + 1) = 88$ (e) $5(n - 1) = 1$ (f) $3(10 + 3m) = 12$

A6 Sadie has started to solve $\frac{n + 3}{4} = 2.5$

Copy and complete her working.

$\frac{n + 3}{4} = 2.5$

$n + 3 = \quad$ (× 4 both sides)

A7 Solve these equations.

(a) $\frac{p - 1}{5} = 3$ (b) $\frac{n + 9}{4} = 1.5$

(c) $\frac{2q + 18}{6} = q$ (d) $\frac{3m - 2}{1.4} = 2m$

A8 Solve these equations.
Think carefully about how to start each time.

(a) $3x + 4 = x - 16$ (b) $3y + 11 = 11y - 5$

(c) $3(z - 2) = 4(z - 5)$ (d) $6(2m - 1) = 60$

(e) $6(2n - 1) = 6n$ (f) $\frac{4p + 5}{3} = 5$

(g) $\frac{q}{2} + 3 = 17$ (h) $\frac{2r - 5}{4} = 3r$

B Solving problems

B1 To work out how long to cook a joint of beef, Tim uses the formula
$$t = 40w + 20$$
where t is the time in minutes and w is the weight in kilograms.

(a) How long will it take to cook a joint of beef that weighs 1.5 kg?
Give your answer in hours and minutes.

(b) Tim wants to find out how much beef he could cook in 3 hours.

(i) Explain why he can do this by solving the equation $40w + 20 = 180$.

(ii) Solve the equation to find how much beef can be cooked in 3 hours.

(c) It takes Tim 2 hours 40 minutes to cook a joint of beef.

Form and solve an equation to find the weight of beef.

B2 Some years ago, temperatures were measured in degrees Fahrenheit (°F).

To convert from degrees Celsius (°C) to °F you can use the formula
$$f = 1.8c + 32$$
where c is the temperature in °C and f is the temperature in °F.

(a) Calculate f when $c = 15$.

(b) Find c when $f = 50$.

(c) Calculate c when $f = 100$, giving the answer to one decimal place.

B3 Pat and Pete think of the same number.

Pat multiplies the number by 3 and adds 4.
Pete subtracts 3 from the number and multiplies by 4.

They both get the same answer.
What was the number?

B4 A square has sides of length l cm.
A triangle has sides of length l cm, $l + 1$ cm and $l + 2$ cm.

(a) If the perimeter of the triangle is 33 cm, what is the area of the square?

(b) Find the value of l if the square and triangle have the same perimeter.

*B5 Before starting work one day, Gareth has £12 and Gina has £1.
Both earn the same amount and at the end of the day, Gareth
has three times as much as Gina.

How much did each earn?

*B6 Cindy is twice as old as her brother Ben.
Three years ago she was five times as old as him.

How old is Ben now?

C Over-balancing?

19 − 3x = 13

    19 = 13 + 3x   (+ 3x both sides)

    6 = 3x   (− 13 both sides)

    2 = x   (÷ 3 both sides)

or x = 2

2(x − 3) = 9 − x

    2x − 6 = 9 − x   (remove brackets)

    3x − 6 = 9   (+ x both sides)

    3x = 15   (+ 6 both sides)

    x = 5   (: 3 both sides)

7 − 5x = 11 − 3x

    7 = 11 + 2x   (+ 5x both sides)

    ⁻4 = 2x   (− 11 both sides)

    ⁻2 = x   (÷ 2 both sides)

or x = ⁻2

C1 Solve the following equations.

(a) $17 - 8x = 1$ (b) $5 - 7x = 40$ (c) $11 = 17 - x$ (d) $4 = 43 - 13x$

C2 Solve the following equations.

(a) $23 - 2k = k + 14$ (b) $2(9 - j) = 19 - 3j$ (c) $11 + h = 18 - h$

(d) $10 - 3g = g + 14$ (e) $4 - 9f = 13f - 29$ (f) $3 + 2e = 3(4 - e)$

(g) $3(4 - d) = 2(3 - d)$ (h) $5 - 7c = 40 - 2c$ (i) $5(2 - b) = 7 - 8b$

(j) $10 + a = 1 - 5a$ (k) $2(5 + c) = 3(5 - c)$ (l) $18 - 4d = 3(6 - d)$

C3 Solve the following equations.

(a) $2.7x - 4 = 5 - 3.3x$ (b) $10 - \frac{1}{2}x = x + 2.5$ (c) $\frac{9 - 3x}{5} = 6$

(d) $\frac{x}{9} = 90 - x$ (e) $2.4(x + 3) = 1.8 - 0.6x$ (f) $\frac{1}{2}x - 5 = 7 - \frac{1}{4}x$

(g) $\frac{x + 2}{7} = 6 - x$ (h) $\frac{5 - 3x}{4} = 5 - 2x$ (i) $\frac{3x + 7.5}{0.5} = x - 5$

C4 Solve the following equations.

(a) $4x - (x + 2) = 25$ (b) $13 - (2x - 3) = 10$ (c) $1 + (3x - 4) = 5 - (x - 2)$

(d) $23 + 2(x - 7) = 15$ (e) $28 - 3(x + 2) = 4$ (f) $3x - 2(x - 5) = 8 - 3(x + 6)$

D More problems

D1 Here are two number puzzles and four equations.

Puzzle 1

I think of a number.

I double it and subtract the result from 12.

The result is four times the number I first thought of.

What number did I think of?

Puzzle 2

I think of a number.

I double it and subtract 12.

The result is four times the number I first thought of.

What number did I think of?

A $12 - 2n = 4$

B $2n - 12 = 4n$

C $12 - 2n = 4n$

D $2n - 12 = 4$

(a) Match each number puzzle to an equation.

(b) Solve each puzzle.

D2 Solve each puzzle by forming an equation and solving it.

(a) I think of a number.
I multiply it by 3 and subtract the result from 21.
My answer is four times the number I first thought of.

What number did I think of?

(b) I think of a number.
I multiply it by 4 and subtract the result from 5.
My answer is 15 more than the number I first thought of.

What number did I think of?

(c) I think of a number.
I subtract it from 2 and multiply the result by 3.
My answer is 2 more than the number I first thought of.

What number did I think of?

(d) I think of a number.
I double it, subtract the result from 3 and then multiply the result by 7.
My answer is the number I first thought of.

What number did I think of?

D3 Alan and Beth are playing a 'think of a number' game.
Each chooses the same starting number.

Alan doubles the number and subtracts the result from 12.
Beth subtracts the number from 12 and then multiplies the result by 3.

They find they both end up with the same result.

(a) What number did they start with?

(b) What number did they end up with?

D4 After Christmas, Adele has £16 and Rick has £9.

Adele buys three zappers while Rick only buys one.
After buying the zappers, they are left with the
same amount of money.

(a) How much is a zapper?

(b) How much money are they each left with?

*D5 At midday, Polly was 5 miles from the crossroads and
Nicola was 8 miles from the same place.

Polly walked towards the crossroads at 2 miles per hour.
Nicola jogged towards the same place at 4 miles per hour.

(a) How far was Polly from the crossroads after x hours?

(b) At what time were they the same distance from the crossroads?

(c) Where was each person at that time?

*D6 People have been interested in puzzles for hundreds of years.

Solve these puzzles from the past.

(a)

'Mother, I wish you would give me a
bicycle,' said a girl of twelve the
other day.

'I do not think you are old enough
yet, my dear,' was the reply. 'When
I am only three times as old as
you are you shall have one.'

Now, the mother's age is forty-five
years. When may the young lady
expect to receive her present?

H. E. Dudeney (1857-1930)

(b)

A carpenter agrees to work on
the condition he is paid 2 groats
for every day he works while he
forfeits 3 groats for every day
he does not work. At the end of
30 days, he finds he has paid out
exactly as much as he has received.

How many days did he work?

Nicholas Chuquet (1484)

(c)

'I say, Rackbrane, what is the time?' an acquaintance
asked our friend the professor the other day. The answer
was certainly curious.

'If you add one quarter of the time from noon till now to
half the time from now till noon tomorrow, you will get
the time exactly.'

What was the time of day when the professor spoke?

H. E. Dudeney (1857-1930)

E Other types of equation

You can often puzzle out a solution to an unfamiliar type of equation.

Examples

Solve $\frac{32}{n} = 4$

> 32 ÷ what = 4 ?
> 32 ÷ 8 = 4, so n = **8**

Solve $\frac{100}{2x^2} = 2$

> 100 ÷ **50** = 2
> So 2x² = 50
> so x² = 25
> so x = **5 or ⁻5**

E1 Solve these equations.

(a) $\frac{12}{x} = 2$

(b) $\frac{1}{n} = 0.5$

(c) $\frac{1}{2p} = \frac{1}{4}$

(d) $\frac{36}{3n} = 4$

(e) $\frac{4}{b} = \frac{1}{4}$

(f) $\frac{18}{n-3} = 2$

(g) $\frac{24}{2y+1} = 3$

(h) $\frac{7}{c} = 14$

(i) $\frac{1}{m} = 10$

E2 Solve these equations.
There may be more than one solution.

(a) $n^2 = 25$

(b) $\frac{a^2}{2} = 32$

(c) $f^2 + 64 = 100$

(d) $\frac{1}{p^2} = \frac{1}{9}$

(e) $3n^2 = 243$

(f) $5n^2 - 2 = 78$

(g) $3(x^2 + 1) = 30$

(h) $x^3 = 8$

(i) $\frac{54}{p^3} = 2$

(j) $\frac{6}{q^2} = 24$

(k) $\frac{5}{n^3} = 5000$

(l) $x^4 = 625$

*E3 Solve these equations, giving all the solutions to each one.

(a) $\frac{x^2 - 5}{10} = 2$

(b) $\frac{12 + x^2}{8} = 6$

(c) $\frac{40 - x^2}{3} = 8$

(d) $\frac{40}{x^2 - 1} = 5$

(e) $7 + \frac{15}{x} = 12$

(f) $20 - \frac{12}{x^2} = 17$

(g) $x(x + 1) = 42$ (This has a positive solution and a negative solution.)

What progress have you made?

Statement	Evidence

I can solve equations like $3(8 + x) = 2x - 1$.

1 Solve the following equations.

   (a) $5v - 4 = 8v - 22$ (b) $2(w + 5) = 15$

   (c) $6(3x + 5) = 12$ (d) $\dfrac{3y - 10}{13} = 2$

   (e) $\dfrac{5z - 2}{4} = z$

2 Solve the following equations.

   (a) $5(p - 1) = 3(p + 3)$

   (b) $5(q + 8) = 15q$

I can solve equations like $3(8 - x) = 2x - 1$.

3 Solve the following equations.

   (a) $24 - 8m = 12$ (b) $15 - 9n = 2n + 37$

   (c) $7(1 - g) = 19 - g$ (d) $\dfrac{4h - 2}{5} = 11 - 3h$

I can solve problems using algebra.

4 Pat and Pete think of the same number.

Pat multiplies the number by 5 and adds 2.
Pete subtracts 8 and multiplies by 2.
They both get the same answer.

What was the number?

5 I think of a number. I multiply it by 6 and subtract the result from 36.
My answer is three times the number I first thought of.

What number did I think of?

6 Adam and Sarah think of the same number.
Adam multiplies the number by 5 and subtracts it from 30.
Sarah subtracts the number from 10 and multiplies by 4.
They both get the same answer.

What was the number?

I can solve other types of simple equation.

7 Solve these.

   (a) $\dfrac{18}{x} = 2$ (b) $n^2 = 16$

   (c) $2n^2 - 3 = 69$ (d) $\dfrac{24}{2n} = 48$

③ Rates

This work will help you

◆ find and interpret the gradient of a straight-line graph

◆ understand and use travel graphs

◆ solve problems involving rates

A Rates of flow

The introductory activity is described in the teacher's guide.

Each graph shows the volume of water coming out of a different tap.

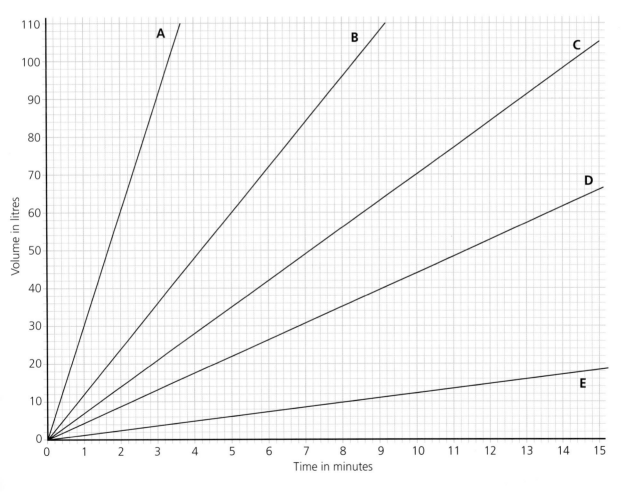

• What is the rate of flow of each tap?

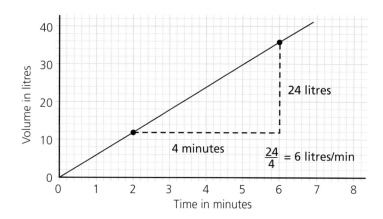

The gradient of the graph (measured using the graph scales) represents the rate of flow.

A1 Find the rate of flow of each of the taps whose graphs are shown here.

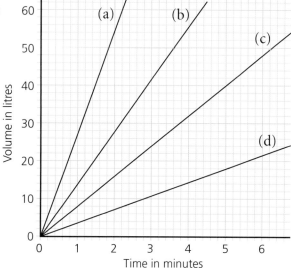

A2 Peter fills a 20 litre bucket from a tap in 4 minutes.

  (a) What is the rate of flow of the tap?

  (b) How long will it take to fill an 85 litre tank from the same tap?

A3 Greta has an indoor and an outdoor tap.
The indoor tap fills a 15 litre bucket in $2\frac{1}{2}$ minutes.
The outdoor tap fills a 10 litre bucket in $1\frac{1}{2}$ minutes.

  (a) Calculate the rate of flow of each tap.

  (b) Calculate how long it will take each tap to fill a 50 litre tank.

A4 Traffic leaving a city has to cross a river.
There are three bridges, whose maximum capacities are 20 vehicles per minute, 25 vehicles per minute and 45 vehicles per minute.

If all three bridges carry vehicles at their maximum rates, how long will it take 1000 vehicles to leave the city?

A5 (a) What do you think is happening in graph A? Find the rate of flow.

(b) Find the rate of flow for graph B.

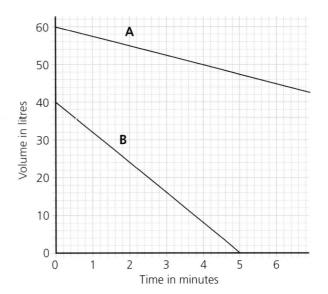

A6 This graph shows the volume of water in Jake's bath.

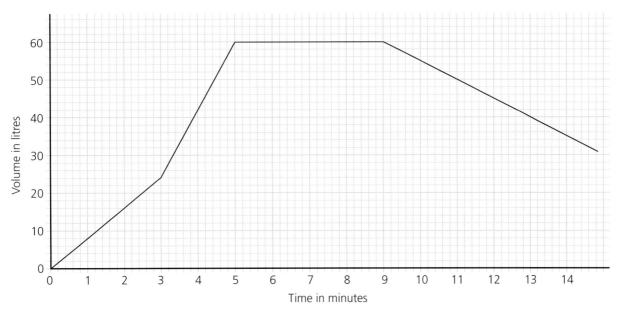

(a) At first Jake turned on the cold tap only. At what rate did the water come out?

(b) Later he turned on the hot tap as well. What was the combined rate of flow of both taps?

(c) What was the rate of flow of the hot tap?

(d) Jake turned both taps off and later pulled out the bath plug. At what rate did the water flow out of the bath?

(e) At what time (measured from when Jake first started filling the bath) will the bath be empty?

B Travel graphs

Donal sets out from home at 10:00.
He drives to the seaside, which is 60 miles away, and gets there at 11:15.

He leaves the seaside at 13:00 and arrives back home at 14:30.
This graph shows his journey.

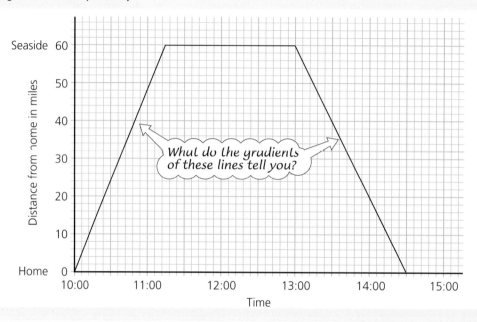

What do the gradients of these lines tell you?

B1 Two cars A and B start together at 8 a.m.
The graphs of their journeys are shown here.

(a) What is the distance between the two cars after $1\frac{1}{2}$ hours?

(b) What is the speed of each car?

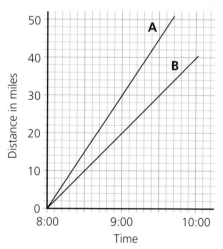

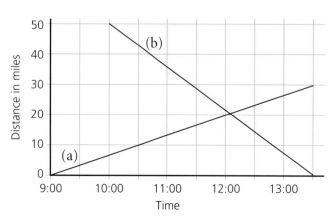

B2 Calculate, to 1 d.p., the speed represented by each graph on the left.

B3 Sara drove to see her aunt, who lives 50 miles from Sara's home. This graph shows her journey there and back.

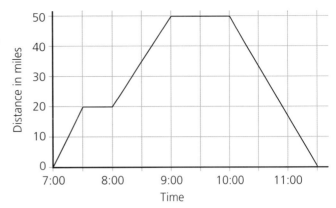

(a) Sara stopped for a break on the way to her aunt's. At what time did she stop and for how long?

(b) At what speed was Sara travelling before she stopped for a break?

(c) At what speed did she travel after her break?

(d) How long did she spend at her aunt's?

(e) At what speed did she travel home?

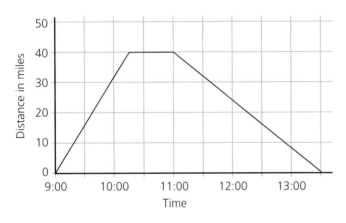

B4 Barbara drove to a friend's house, stayed there for a while and then drove back. Her journey is shown in this graph.

(a) At what speed did she drive to her friend's house?

(b) How long did she spend there?

(c) At what speed did she drive back?

B5 Justin drove from his home to a holiday camp, 90 miles away. He set out at 08:00 and drove at 40 m.p.h. all the way.

Justin's wife, Mary, also drove from home to the camp. She set out at 08:30 and arrived at 10:00.

(a) Draw both of their travel graphs on the same axes.

(b) How fast did Mary drive?

(c) Where and when did Mary overtake Justin?

*B6 A wagon train is lumbering through the desert at 3 m.p.h. At 08:00 a scout on a pony leaves the wagon train and rides ahead at 15 m.p.h.

At 10:30 the scout sees a hostile force, turns round and rides back to the wagon train at 15 m.p.h.

By drawing a travel graph, find the time at which the scout gets back to the wagon train.

C Problems involving rates

Problems involving rates usually lead to multiplication or division.
Often the difficulty is deciding which to do.
It sometimes helps to simplify the numbers.

Example   Water leaks from a tank at a rate of 0.45 litre per hour.
The tank holds 7.38 litres when full.
How long will it take to empty?

Make up a similar question with 'easy' whole numbers.   *Every 3 litres takes an hour to leak away.*
*Suppose the water leaks at **3 litre per hour** and the*   *So for 12 litres it will take **4 hours**.*
*tank holds **12 litres**.*   *You get the answer by **dividing 12 by 3**.*

So in the original problem you must divide 7.38 by 0.45. $\dfrac{7.38}{0.45}$ = 16.4 hours

C1 A tap flows at a rate of 0.8 litre per second.

(a) How much water comes out in 12.5 seconds?

(b) How long will it take to fill a bucket of capacity 20 litres?

C2 Calculate the missing quantity in each of these.

(a) Rate = 3.2 litre per second, time = 45 seconds, volume = ?

(b) Time = 20.5 seconds, volume = 13 litres, rate = ?

(c) Rate = 0.16 litre per second, volume = 27.8 litres, time = ?

C3 A tap fills a 21.6 litre container in 4.5 seconds.

(a) What is the rate of flow of the tap in litres per second?

(b) How much water comes out of the tap in 20 seconds?

(c) How long (to the nearest second) will the tap take to fill a 450 litre container?

C4 The water in a reservoir is draining away.
The volume of water went down from 45 000 litres to 41 500 litres in 4 days.
After how many more days will the reservoir be empty?

C5 Karl's printer took 32 minutes to print 144 pages.
Julie's printer took 21 minutes to print 90 pages.
How long will the faster printer take to print 60 pages?

C6 A bottle-filling machine fills 14 bottles per minute.
An improvement to the machine is suggested that will increase the rate to 18 bottles per minute.

How much less time will the improved machine take to fill 5000 bottles?
Give your answer to a reasonable degree of accuracy.

C7 The population density of a region can be measured in **persons per km²** by dividing the population of the region by the area in km².

    (a) A region X of area 22.5 km² has a population of 141 300.
        Calculate its population density.

    (b) Region X is next to another region Y.
        Region Y has a population density of 4160 persons per km² and an area of 31.5 km².

        Regions X and Y are to be combined to form a single region.
        Calculate the population density of the single region, to three significant figures.

*C8 Jane's bath has a capacity of 240 litres.
She turns on the hot tap alone and it fills the bath in 25 minutes.
Next time she uses the cold tap alone and it fills the bath in 16 minutes.

How long will the bath take to fill if both taps are on?
(The rate of flow of each tap is unaffected by whether the other is on or off.)

What progress have you made?

Statement

I can find a rate from a graph.

Evidence

1 What rate of flow does each graph show?

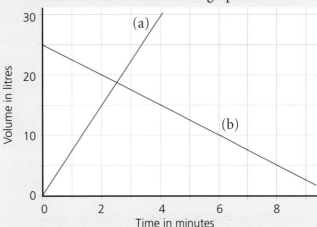

I can understand and use a travel graph.

2 Paul joins a motorway at 09:00 and drives at a steady speed of 40 m.p.h. Sandra joins the motorway at the same point as Paul but at 09:45. She drives at a steady speed of 60 m.p.h.

Draw their travel graphs on the same axes and find when and where Sandra overtakes Paul.

I can solve problems involving rates.

3 Water leaks from a tank at a rate of 0.65 litre per minute. At present there are 2.6 litres of water in the tank.

After how long will the tank be empty?

27

④ Vectors

This is about movements between points on a grid.
The work will help you

◆ understand a way of writing these movements (vectors)

◆ solve simple problems involving more than one vector

◆ understand what adding vectors means

A Writing and drawing vectors

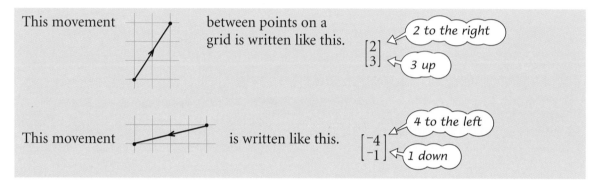

This movement ⬚ between points on a grid is written like this.

$\begin{bmatrix} 2 \\ 3 \end{bmatrix}$

2 to the right

3 up

This movement ⬚ is written like this.

$\begin{bmatrix} -4 \\ -1 \end{bmatrix}$

4 to the left

1 down

How will these be written?

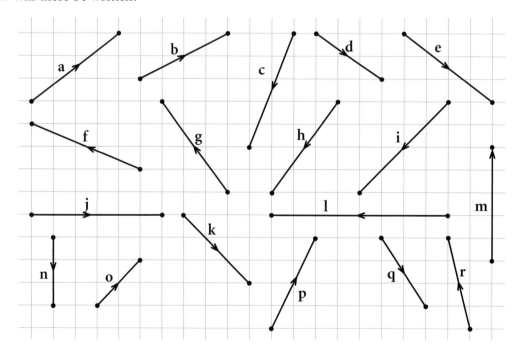

28

Vector snakes and ladders   a game for two players

The vector cards are made from sheet 247 and the game board is sheet 246.

Basic game

Before you start

Put the pile of cards face down on the table.

Put both counters on START.

When it is your turn

Turn over the top card. This tells you how to move your counter.
(Your counter always goes where two gridlines cross,
not in a square.)

If the move would take you off the board, you miss that go and
your partner can either use that move on their go or
turn over the next top card.

You slide down snakes and go up ladders in the usual way.

Strategy game

Before you start

Deal four cards to each player and put the rest in a pile face down.

Put both counters on START.

When it is your turn

Choose one of your cards to decide how your counter moves.
When you have used a card, put it at the bottom of the pack
and take a new card from the top.

If you wish, you may replace one of your cards with
one from the top of the pack **without moving**.

A1 On squared paper, draw arrows to show these movements.

(a) $\begin{bmatrix} -1 \\ 5 \end{bmatrix}$ (b) $\begin{bmatrix} -2 \\ -2 \end{bmatrix}$ (c) $\begin{bmatrix} 2 \\ 5 \end{bmatrix}$ (d) $\begin{bmatrix} -5 \\ 0 \end{bmatrix}$ (e) $\begin{bmatrix} 3 \\ 4 \end{bmatrix}$ (f) $\begin{bmatrix} 1 \\ -4 \end{bmatrix}$ (g) $\begin{bmatrix} -2.5 \\ 3.5 \end{bmatrix}$ (h) $\begin{bmatrix} 0 \\ -3 \end{bmatrix}$

A2 A vector written as one number above another in brackets is called a **column vector**.
Write these as column vectors.

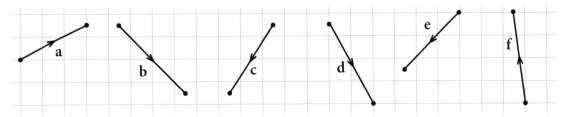

29

B Combining vectors

B1 Mark a point P on some squared paper.

(a) Draw a journey consisting of these vectors.

$$\begin{bmatrix} 3 \\ 0 \end{bmatrix}, \text{ then } \begin{bmatrix} 2 \\ 1 \end{bmatrix}, \text{ then } \begin{bmatrix} 1 \\ 3 \end{bmatrix}$$

Mark the finishing point Q.

(b) Start again at the point P you marked.
This time draw the vectors in this order.

$$\begin{bmatrix} 1 \\ 3 \end{bmatrix}, \qquad \begin{bmatrix} 2 \\ 1 \end{bmatrix}, \qquad \begin{bmatrix} 3 \\ 0 \end{bmatrix}$$

Where do you finish?

(c) What happens when you start from P and
use the vectors in this order?

$$\begin{bmatrix} 2 \\ 1 \end{bmatrix}, \qquad \begin{bmatrix} 3 \\ 0 \end{bmatrix}, \qquad \begin{bmatrix} 1 \\ 3 \end{bmatrix}$$

B2 Mark points M and N carefully on some squared paper.
You have to draw a journey from M to N.

You must use $\begin{bmatrix} 4 \\ 3 \end{bmatrix}$, $\begin{bmatrix} 1 \\ -2 \end{bmatrix}$ and one other vector.

What is this other vector?

B3 You want to get from R to S.
You can use three out of these four vectors.

$$\begin{bmatrix} 4 \\ 2 \end{bmatrix}, \begin{bmatrix} 2 \\ 2 \end{bmatrix}, \begin{bmatrix} 2 \\ 3 \end{bmatrix}, \begin{bmatrix} 3 \\ -1 \end{bmatrix}$$

(a) Which three vectors do you need?

(b) Does it matter in which order you use them?

B4 Choose three out of these five vectors
to get you from A to B.

$$\begin{bmatrix} 2 \\ 3 \end{bmatrix}, \begin{bmatrix} 4 \\ 2 \end{bmatrix}, \begin{bmatrix} -1 \\ 2 \end{bmatrix}, \begin{bmatrix} 7 \\ -1 \end{bmatrix}, \begin{bmatrix} 4 \\ 1 \end{bmatrix}$$

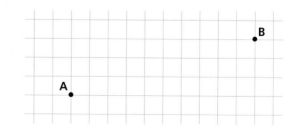

B5 (a) You want to get from C to D.

Use two out of these four vectors.

$$\begin{bmatrix}1\\-2\end{bmatrix}, \begin{bmatrix}3\\2\end{bmatrix}, \begin{bmatrix}4\\1\end{bmatrix}, \begin{bmatrix}2\\-2\end{bmatrix}$$

C•

•D

(b) You can get from C to D using three out of these five vectors.

Which ones do you need?

$$\begin{bmatrix}-1\\-2\end{bmatrix} \quad \begin{bmatrix}3\\1\end{bmatrix} \quad \begin{bmatrix}-1\\1\end{bmatrix} \quad \begin{bmatrix}-2\\-2\end{bmatrix} \quad \begin{bmatrix}7\\0\end{bmatrix}$$

A journey made from several separate vectors can be written as an addition.

$$\begin{bmatrix}3\\7\end{bmatrix} + \begin{bmatrix}-1\\3\end{bmatrix} + \begin{bmatrix}5\\-2\end{bmatrix} = \begin{bmatrix}7\\8\end{bmatrix}$$

This result is the vector that would take you straight from the starting point to the finishing point.

B6 Work out an answer for each of these without doing a drawing.

(a) $\begin{bmatrix}2\\6\end{bmatrix} + \begin{bmatrix}0\\4\end{bmatrix} + \begin{bmatrix}3\\-1\end{bmatrix}$ (b) $\begin{bmatrix}6\\0\end{bmatrix} + \begin{bmatrix}-1\\4\end{bmatrix} + \begin{bmatrix}-3\\1\end{bmatrix}$ (c) $\begin{bmatrix}2\\-4\end{bmatrix} + \begin{bmatrix}3\\6\end{bmatrix} + \begin{bmatrix}0\\1\end{bmatrix}$

(d) $\begin{bmatrix}-1\\3\end{bmatrix} + \begin{bmatrix}2\\2\end{bmatrix} + \begin{bmatrix}3\\-6\end{bmatrix}$ (e) $\begin{bmatrix}-3\\5\end{bmatrix} + \begin{bmatrix}-1\\6\end{bmatrix} + \begin{bmatrix}2\\-1\end{bmatrix}$ (f) $\begin{bmatrix}2\\-1\end{bmatrix} + \begin{bmatrix}1\\-6\end{bmatrix} + \begin{bmatrix}5\\-2\end{bmatrix}$

B7 Work out the vector additions using these vectors.

(a) $\mathbf{p} + \mathbf{q}$ (b) $\mathbf{q} + \mathbf{r}$ (c) $\mathbf{q} + \mathbf{s} + \mathbf{u}$

(d) $\mathbf{r} + \mathbf{s}$ (e) $\mathbf{s} + \mathbf{t}$ (f) $\mathbf{p} + \mathbf{r} + \mathbf{u}$

(g) $\mathbf{t} + \mathbf{u}$ (h) $\mathbf{p} + \mathbf{u}$ (i) $\mathbf{r} + \mathbf{s} + \mathbf{t}$

$$\mathbf{p} = \begin{bmatrix}8\\7\end{bmatrix} \quad \mathbf{q} = \begin{bmatrix}3\\-4\end{bmatrix} \quad \mathbf{r} = \begin{bmatrix}0\\6\end{bmatrix}$$

$$\mathbf{s} = \begin{bmatrix}-1\\5\end{bmatrix} \quad \mathbf{t} = \begin{bmatrix}-3\\-2\end{bmatrix} \quad \mathbf{u} = \begin{bmatrix}9\\-4\end{bmatrix}$$

B8 Find the missing numbers in these column vectors.

$$\begin{bmatrix}2\\a\end{bmatrix} + \begin{bmatrix}b\\-2\end{bmatrix} = \begin{bmatrix}7\\-1\end{bmatrix} \qquad \begin{bmatrix}6\\c\end{bmatrix} + \begin{bmatrix}d\\-2\end{bmatrix} = \begin{bmatrix}6\\4\end{bmatrix} \qquad \begin{bmatrix}4\\-1\end{bmatrix} + \begin{bmatrix}-7\\e\end{bmatrix} + \begin{bmatrix}f\\4\end{bmatrix} = \begin{bmatrix}-1\\4\end{bmatrix}$$

$$\begin{bmatrix}5\\g\end{bmatrix} + \begin{bmatrix}h\\-3\end{bmatrix} = \begin{bmatrix}1\\-7\end{bmatrix} \qquad \begin{bmatrix}2\\i\end{bmatrix} + \begin{bmatrix}j\\5\end{bmatrix} = \begin{bmatrix}-6\\8\end{bmatrix} \qquad \begin{bmatrix}1\\k\end{bmatrix} + \begin{bmatrix}l\\-3\end{bmatrix} + \begin{bmatrix}6\\1\end{bmatrix} = \begin{bmatrix}6\\-8\end{bmatrix}$$

B9 Find the missing column vectors.

(a) $\begin{bmatrix}8\\2\end{bmatrix} + \begin{bmatrix}?\end{bmatrix} = \begin{bmatrix}4\\3\end{bmatrix}$ (b) $\begin{bmatrix}?\end{bmatrix} + \begin{bmatrix}-4\\-1\end{bmatrix} = \begin{bmatrix}-7\\2\end{bmatrix}$ (c) $\begin{bmatrix}-3\\2\end{bmatrix} + \begin{bmatrix}9\\7\end{bmatrix} + \begin{bmatrix}?\end{bmatrix} = \begin{bmatrix}1\\8\end{bmatrix}$

(d) $\begin{bmatrix}?\end{bmatrix} + \begin{bmatrix}-1\\4\end{bmatrix} = \begin{bmatrix}2\\-6\end{bmatrix}$ (e) $\begin{bmatrix}8\\6\end{bmatrix} + \begin{bmatrix}?\end{bmatrix} = \begin{bmatrix}-4\\8\end{bmatrix}$ (f) $\begin{bmatrix}?\end{bmatrix} + \begin{bmatrix}-7\\4\end{bmatrix} + \begin{bmatrix}1\\-2\end{bmatrix} = \begin{bmatrix}8\\-3\end{bmatrix}$

B10 Work out the column vectors that the letters stand for.

(a) $\begin{bmatrix} 4 \\ 2 \end{bmatrix} + \mathbf{j} = \begin{bmatrix} 7 \\ 6 \end{bmatrix}$ (b) $\mathbf{k} + \begin{bmatrix} 2 \\ 7 \end{bmatrix} = \begin{bmatrix} 1 \\ 9 \end{bmatrix}$ (c) $\begin{bmatrix} 4 \\ 4 \end{bmatrix} + \mathbf{l} = \begin{bmatrix} ^-1 \\ ^-2 \end{bmatrix}$

(d) $\begin{bmatrix} 4 \\ 5 \end{bmatrix} + \mathbf{m} + \mathbf{m} = \begin{bmatrix} 8 \\ 7 \end{bmatrix}$ (e) $\mathbf{n} + \mathbf{n} + \begin{bmatrix} 10 \\ ^-8 \end{bmatrix} = \begin{bmatrix} 6 \\ ^-2 \end{bmatrix}$ (f) $\begin{bmatrix} 2 \\ ^-2 \end{bmatrix} + \mathbf{o} + \mathbf{o} + \mathbf{o} = \begin{bmatrix} ^-7 \\ 7 \end{bmatrix}$

Round trip challenge

1 On squared paper, draw arrows for these vectors.

$\mathbf{a} = \begin{bmatrix} ^-2 \\ 4 \end{bmatrix}$ $\mathbf{b} = \begin{bmatrix} 4 \\ 0 \end{bmatrix}$ $\mathbf{c} = \begin{bmatrix} 2 \\ 4 \end{bmatrix}$ $\mathbf{d} = \begin{bmatrix} ^-4 \\ 2 \end{bmatrix}$ $\mathbf{e} = \begin{bmatrix} 0 \\ ^-2 \end{bmatrix}$ $\mathbf{f} = \begin{bmatrix} 4 \\ ^-2 \end{bmatrix}$

$\mathbf{g} = \begin{bmatrix} ^-2 \\ ^-2 \end{bmatrix}$ $\mathbf{h} = \begin{bmatrix} ^-2 \\ ^-4 \end{bmatrix}$ $\mathbf{i} = \begin{bmatrix} 4 \\ 8 \end{bmatrix}$ $\mathbf{j} = \begin{bmatrix} 2 \\ ^-4 \end{bmatrix}$ $\mathbf{k} = \begin{bmatrix} ^-4 \\ 0 \end{bmatrix}$

Label your arrows with their letters.
Check your drawings carefully.

2 If you start at point P and do a journey consisting of **a** then **f** then **g**,
you get back to the starting point.

How can you tell this will happen, just by looking at the column vectors?

Draw the journey. What special type of triangle do you get?

3 Choose three of the vectors to draw a 'round trip' that is
a different triangle of this special type.
Label the vector arrows with their letters.

4 Choose three of the vectors to draw a round trip that is a right-angled triangle.
Label the vector arrows with their letters.

5 If you do a journey consisting of **a** then **f** then **e**,
what new column vector do you need to get back to the starting point?
What kind of shape does this make?

6 Choose four of the vectors to draw a round trip that is a square.
Label the vector arrows with their letters.

7 In the same way, draw and label

(a) two different rhombuses

(b) two different parallelograms

(c) two different trapeziums

What progress have you made?

Statement

I know how to write and draw vectors.

Evidence

1 Draw these vectors on squared paper.

(a) $\begin{bmatrix} 2 \\ 1 \end{bmatrix}$ (b) $\begin{bmatrix} 3 \\ -4 \end{bmatrix}$ (c) $\begin{bmatrix} -2 \\ 4 \end{bmatrix}$

(d) $\begin{bmatrix} -1 \\ -6 \end{bmatrix}$ (e) $\begin{bmatrix} 0 \\ 3 \end{bmatrix}$ (f) $\begin{bmatrix} -6 \\ 0 \end{bmatrix}$

2 Write these vectors.

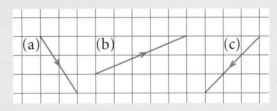

I can solve simple problems involving more than one vector.

3 You need to go from A to B using just two vectors.

Give three possible ways of doing this.

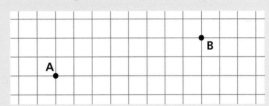

I can work with vector additions.

4 Work out $\begin{bmatrix} 3 \\ 7 \end{bmatrix} + \begin{bmatrix} -1 \\ 5 \end{bmatrix} + \begin{bmatrix} 0 \\ -6 \end{bmatrix}$.

5 Find the missing column vectors.

(a) $\begin{bmatrix} 2 \\ -4 \end{bmatrix} + \begin{bmatrix} ? \\ ? \end{bmatrix} + \begin{bmatrix} -8 \\ 2 \end{bmatrix} = \begin{bmatrix} -1 \\ 0 \end{bmatrix}$

(b) $\begin{bmatrix} -3 \\ 8 \end{bmatrix} + \begin{bmatrix} 7 \\ 0 \end{bmatrix} + \begin{bmatrix} ? \\ ? \end{bmatrix} = \begin{bmatrix} 5 \\ 16 \end{bmatrix}$

⑤ Manipulation

This work will help you

◆ simplify a variety of algebraic expressions

◆ use algebra to solve problems and give explanations

A Multiplying

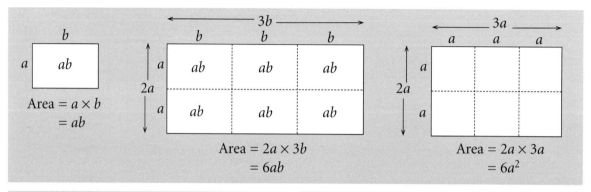

$$5d^2 \times 3d = 5 \times d \times d \times 3 \times d$$
$$= 5 \times 3 \times d \times d \times d$$
$$= 15d^3$$

$$2a \times 5ab = 2 \times a \times 5 \times a \times b$$
$$= 2 \times 5 \times a \times a \times b$$
$$= 10a^2b$$

A1 Find expressions for the areas of these rectangles.
Write each expression in its simplest form.

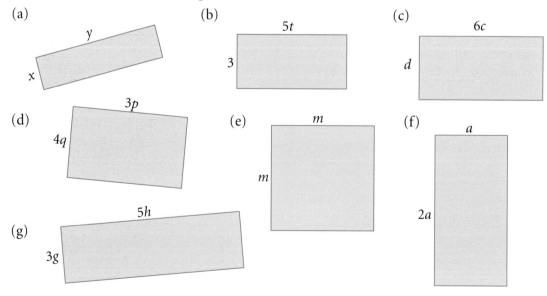

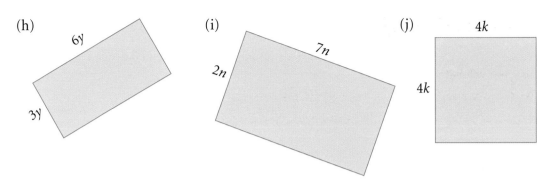

(h)

6y

3y

(i)

7n

2n

(j)

4k

4k

A2 Find the result of each multiplication in its simplest form.

(a) $4g \times h$

(b) $8m \times 7$

(c) $x \times 7x$

(d) $2a \times 4b$

(e) $5c \times d$

(f) $e \times 3f$

(g) $3g \times 6h$

(h) $2k \times 3k$

(i) $6a^2 \times a$

(j) $2b^2 \times 5c$

(k) $3d^2 \times 2d$

(l) $2ef \times 3f$

A3 Find the result of each multiplication in its simplest form.

(a) $h \times h \times 2h$

(b) $8b^3 \times 3$

(c) $5t \times 6s^3$

(d) $3g \times 2g^2$

(e) $5h \times 3jh$

(f) $3m \times 2mn$

(g) $2ab \times 3ab$

(h) $4c^2 \times 2cd$

(i) $fg \times 3fg^2$

Cover up

You need sheet 248.

Cut out the 18 rectangular pieces.

Put the pieces on the board so that each piece covers a pair of expressions which multiply to give that answer.

For example, $6ab$ may cover $6a$ and b, or it may cover $2b$ and $3a$.

The pieces can be put

this way ☐ or this way ☐ .

$6a$	b	$8b^2$	a	a^2	
a	$4a$	$2a$	$4b$	$4a$	$2b$
$3a^2$	b	$2a$	$4b^2$	$2b$	
$4ab$	2	$3b$	5	$2a$	$3a$
ab^2					
6					

$12ab$

$6a^2b^2$

$4a^2b^3$

$6ab$

$6a^2$

$8b^2$

$8ab$

$4ab$

$8b^3$

$10a$

x^2b

$4a^3$

$2a^3$

$6a^2b$

$9a^2$

$4ab^2$

- How many pieces can you put on the board?
- Can you find a way to cover the whole board?

B Dividing

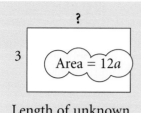

Length of unknown
side is $\frac{12a}{3} = 4a$.

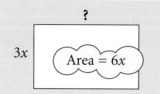

Length of unknown
side is $\frac{6x}{3x} = 2$.

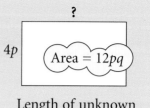

Length of unknown
side is $\frac{12pq}{4p} = 3q$.

B1 Find expressions for the unknown sides in these rectangles.

(a)

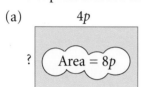

(b)

$$Area = 16q$$

(c)

$$m$$
$$Area = mn$$

(d)

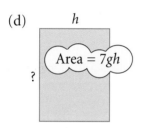

(e)

$$5a$$
$$Area = 10ab$$

(f)

$$6y$$
$$Area = 12xy$$

(g)

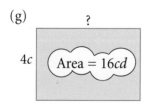

(h)

$$x$$
$$Area = x^2$$

(i)

$$2a$$
$$Area = 2a^2$$

(j)

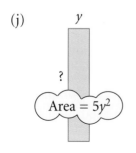

(k)

$$2k$$
$$Area = 6k^2$$

(l)

$$8p$$
$$Area = 4p^2$$

B2 Find an expression for the area of each triangle.

(a)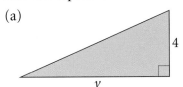

(b) (c)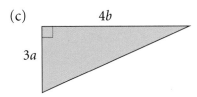

B3 Find an expression for the length of each unknown side marked '?'.

(a)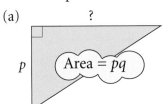

(b)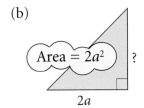

(c)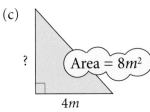

B4 Simplify these.

(a) $\dfrac{15n}{5}$

(b) $\dfrac{7m}{m}$

(c) $\dfrac{4jk}{j}$

(d) $\dfrac{3h^2}{h}$

(e) $\dfrac{6fg}{3g}$

(f) $\dfrac{8e^2}{2e}$

(g) $\dfrac{10cd}{5cd}$

(h) $\dfrac{12ab^2}{2b}$

You can simplify fractions by dividing the numerator and denominator by a common factor.

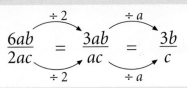 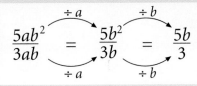

B5 Simplify these.

(a) $\dfrac{6z}{4}$

(b) $\dfrac{5xy}{4y}$

(c) $\dfrac{9v}{6w}$

(d) $\dfrac{2u^2}{3u}$

(e) $\dfrac{8rs}{2rt}$

(f) $\dfrac{3pq^2}{2pq}$

(g) $\dfrac{5n^2}{3n^2}$

(h) $\dfrac{12km}{9k}$

(i) $\dfrac{7gj}{gh}$

(j) $\dfrac{6e}{3ef}$

(k) $\dfrac{10cd^2}{3cd}$

(l) $\dfrac{3ab}{9a^2b}$

B6 Solve the puzzle on sheet 249.

C Brackets

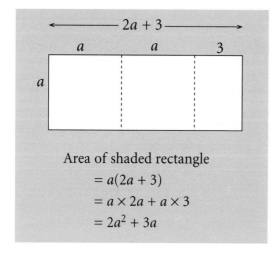

Area of shaded rectangle
$$= a(2a + 3)$$
$$= a \times 2a + a \times 3$$
$$= 2a^2 + 3a$$

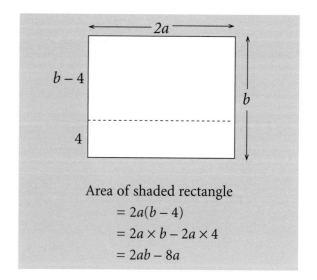

Area of shaded rectangle
$$= 2a(b - 4)$$
$$= 2a \times b - 2a \times 4$$
$$= 2ab - 8a$$

C1 For each shaded rectangle, write an expression for the area
- with brackets – for example, $a(2a + 3)$
- without brackets – for example, $2a^2 + 3a$

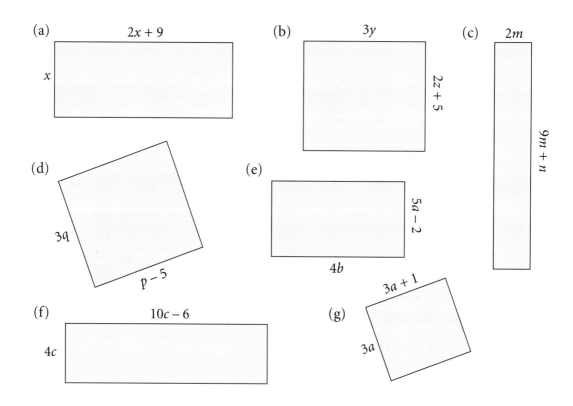

C2 Find four pairs of equivalent expressions.

A $2a(a + 4b)$

B $2(a^2 + 2b)$

C $3a(4b + 9a)$

D $2a^2 + 4b$

E $12ab + 27a^2$

F $3a(4b + 3)$

G $2a^2 + 8ab$

H $12ab + 9a$

C3 Multiply out the brackets from each expression.

(a) $x(2x + 4)$

(b) $y(y - 5)$

(c) $5(3x - y)$

(d) $x(6 - x)$

(e) $x(y + z)$

(f) $2x(3y + 1)$

(g) $x(5x - 6)$

(h) $2x(3x - 5y)$

(i) $4y(3z + 2y)$

(j) $2x(xy + z^2)$

(k) $xy(x - y)$

(l) $2xy(x + 3y)$

C4 Find expressions for the unknown sides in these rectangles.

(a) ?
m Area $= m^2 + 4m$

(b) $k - 6$
? Area $= hk - 6h$

(c) ?
$2e$ Area $= 2ef + 2eg$

(d) ?
$3d$ Area $= 3d^2 - 3d$

(e) $3c + 5$
? Area $= 6c^2 + 10c$

(f) ?
$5a$ Area $= 15a^2 - 10ab$

C5 Find the missing expressions to make these statements correct.

(a) $\square(2s - 5) = 2s^2 - 5s$

(b) $\square(r + 6) = 2r^2 + \square$

(c) $6q(q - \square) = \square - 12pq$

(d) $\square(2m + n) = \square + nt$

(e) $\square(2k - 1) = 8k^2 - \square$

(f) $hj(2 + \square) = \square + hj^2$

(g) $5f(\square + \square) = 10fg + 5f$

(h) $\square(3e^2 - 1) = \square - 2d$

(i) $\square(\square + \square) = c^2 + 5c$

(j) $\square(\square - \square) = 7a^2 - 7b$

D True to form

In each statement, n is a positive integer.

Which are
- always true?
- sometimes true?
- never true?

A 4n is a multiple of 4

B 2n + 5 is odd

C n + 10 is negative

D 5n + 1 is a multiple of 5

E 4n is a multiple of 8

F 6(n + 1) is a multiple of 6

G 2n is odd

H n² is a multiple of 5

I 6n + 8 is even

J 1 – n is negative

K (3n)² is a multiple of 9

L 3n + 6 is a multiple of 3

D1 In each expression, n is a positive integer.

A $3n + 1$

B $5n + 10$

C $12n$

D $8n - 4$

E $(2n)^2$

F $4n + 1$

G $2n^2$

H $\dfrac{6n^2}{3n}$

I $3n + 6$

J $6n + 9$

K $5(2n - 1)$

L $25n^2$

List the expressions that are always
(a) even
(b) a square number
(c) a multiple of 3
(d) a multiple of 4
(e) odd
(f) a multiple of 5
(g) one more than a multiple of 3

D2 Write down an expression for a number that is always a multiple of 2 **and** a multiple of 3.

E Grid totals

This grid of numbers has ten columns.

A T-shape outlines some numbers.

1	2	3	4	5	6	7	8	9	10
11	12	1.	14	15	16	17	18	19	20
21	22	23		25	20	27	28	29	30
31	32	33	3-	35	36	37	38	39	40
		43	44	45	46	47	48		

E1 (a) What is the total of the numbers in the T-shape?

   (b) Find totals for the T-shape in different positions on the grid.
       What can you say about your totals?

   (c) (i) Copy and complete this
           T-shape for the grid above.

       (ii) Find an expression for the total.

       (iii) What does this expression show about
            T-shape totals on this grid?

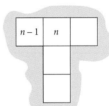

   (d) Which position of the T-shape will give a total of 200?

E2 Investigate totals for the following shapes.

   (a)   (b)   (c)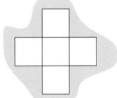

E3 Draw a shape that always gives a multiple of 6 as its total.

E4 Another grid of numbers
   has six columns.

1	2	3	4	5	6
7	8	9	10	11	12
13	14	15	16	17	18
	20	21	2.		
		28			

   (a) (i) Copy and complete this T-shape
           for the six-column grid.

       (ii) Find an expression for the total.

   (b) Investigate T-shape totals for grids with different numbers of columns.

   (c) Can you find an expression for the T-shape total for a grid with m columns?

41

F Ways of seeing

Joy works in a bakery.
One way she decorates square cakes is shown below.

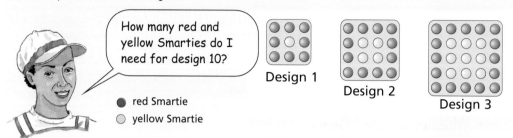

How many red and yellow Smarties do I need for design 10?

- ● red Smartie
- ○ yellow Smartie

Design 1

Design 2

Design 3

F1 Here are some triangular cakes.

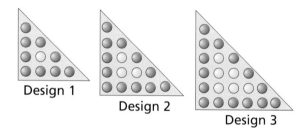

Design 1

Design 2

Design 3

(a) How many red Smarties would be in design 6?

The diagrams below show how some pupils counted the red Smarties in design 5.

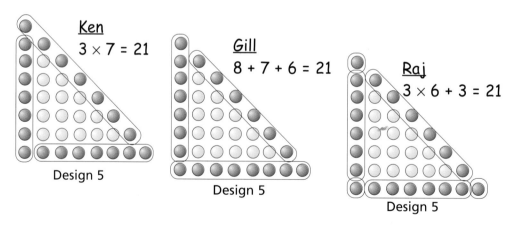

Ken
$3 \times 7 = 21$

Design 5

Gill
$8 + 7 + 6 = 21$

Design 5

Raj
$3 \times 6 + 3 = 21$

Design 5

(b) How do you think each pupil would count the red Smarties in design 7?

(c) They each found a rule for the number of red Smarties in design n.

$$r = (n + 3) + (n + 2) + (n + 1) \qquad r = 3(n + 1) + 3 \qquad r = 3(n + 2)$$

(i) Which of these rules are correct?

(ii) Who do you think found each rule?
Explain your answers carefully.

F2 Here are some rectangular cakes.

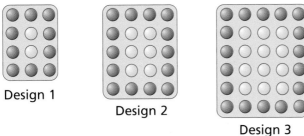

Design 1

Design 2

Design 3

(a) Draw a diagram to show design 4.

(b) For design 4, what is the number of

    (i) yellow Smarties

    (ii) red Smarties

(c) How many red and yellow Smarties would you need for design 10?

(d) Find an expression for the number of yellow Smarties in design n.

(e) Find an expression for the number of red Smarties in design n. Write your expression in its simplest form.

(f) How many red Smarties would you need for design 100?

F3 Here is another design for square cakes.

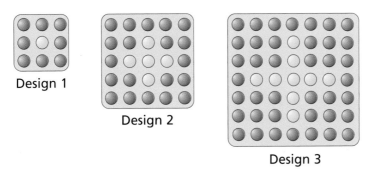

Design 1

Design 2

Design 3

(a) For design 3, what is the number of

    (i) yellow Smarties

    (ii) red Smarties

(b) How many red and yellow Smarties would you need for design 15?

(c) Find an expression for the number of yellow Smarties in design n.

(d) Find an expression for the number of red Smarties in design n.

(e) Show that the number of red Smarties is always a multiple of 4.

G Ways of seeing further

T

* What about a stack like this with n rows of tins?

G1 These diagrams show another way to stack tins.

(a) How many tins are in the stack with 4 rows of tins?

(b) How many tins are in a stack like this with 6 rows?

(c) Find an expression for the number of tins in a stack with n rows.

G2 The diagrams below show how to draw a 'mystic rose'.

This is a
7-point
mystic rose.

Mark equally spaced
points round a circle.

Join the points with
straight lines.

Continue until no more
lines are possible.

(a) Draw three different mystic roses.
    How many straight lines are in each of your designs?

(b) How many lines would be in a 20-point mystic rose?

(c) Find an expression for the number of lines in a n-point mystic rose.

G3 (a) A group of three people shake hands with each other.
How many handshakes take place?

(b) Investigate the number of handshakes for different groups of people.

(c) What would be the total number of handshakes for 100 people?

(d) Find an expression that gives the total number of handshakes for n people.

G4 Below is a drawing of a 'step tower'.

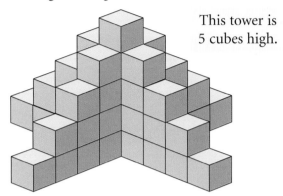

This tower is
5 cubes high.

(a) How many cubes are in this step tower?

(b) How many cubes would you need to make a tower

(i) 3 cubes high

(ii) 15 cubes high

(c) Find an expression that gives the total
number of cubes for a tower n cubes high.

Explain clearly how you found your expression.

*G5** This diagram shows strands of wire inside a cable.
The strands of wire form a hexagon.

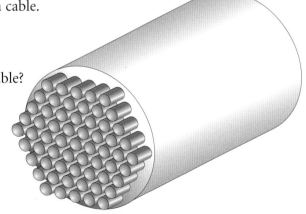

(a) Why do you think it is called a size 5 cable?

(b) What is the total number of
strands inside this cable?

(c) Draw diagrams and count the
number of strands of wire in

(i) a size 4 cable

(ii) a size 7 cable

(d) How many strands of wire would be inside a size 20 cable?

(e) Find an expression for the number of strands in a size n cable.

Explain clearly how you found your expression.

45

What progress have you made?

Statement

I can multiply and divide expressions.

Evidence

1 Write the result of each multiplication in its simplest form.

  (a) $3a \times 5b$ (b) $2k \times 4k$ (c) $3p^2 \times 4q$

  (d) $5r \times 3r^2$ (e) $3x \times 3xy$

2 Find expressions for the unknown sides in these rectangles.

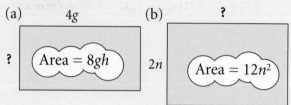

3 Simplify these.

  (a) $\dfrac{6d^2}{2d}$ (b) $\dfrac{5gh^2}{15gh}$ (c) $\dfrac{12x}{8x^2}$

I can multiply out brackets.

4 Multiply out the brackets.

  (a) $3x(y - z)$ (b) $3m(2n + 3m)$

I can use algebra to help solve problems and to carry out mathematical investigations.

5 This grid of numbers has five columns. A square shape outlines some numbers.

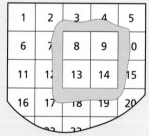

  (a) Find totals for the square shape in different positions on the grid.

  (b) Use algebra to show the total will always be a multiple of 4.

  (c) What numbers are in the square shape if the total is 220?

All your work in sections D, E, F and G is also evidence of this.

6 Area of a circle

This work will help you

◆ calculate the area of a circle from its diameter or radius

◆ solve problems that involve circumference or area

◆ find the circumference and area in terms of π

◆ find the volume and surface area of a cylinder

A The formula for the area of a circle

• Explain why diagram A shows that the area of the circle is less than 4 times r^2.

• Explain why diagram B shows that the area of the circle is more than 2 times r^2.

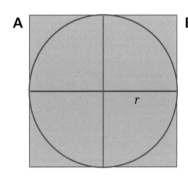
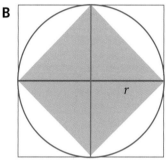

The areas of the circles below have been measured approximately by counting squares.

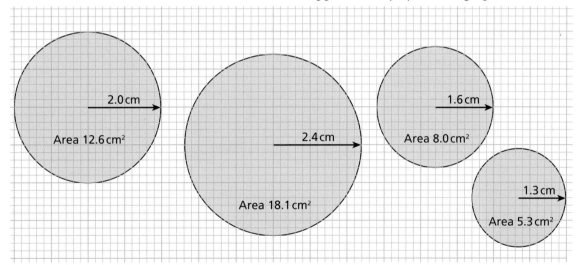

• Make a table showing r^2 and the area A for each circle. Work out how many times r^2 goes into A.

What do you think is the formula connecting A and r?

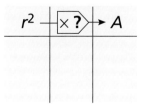

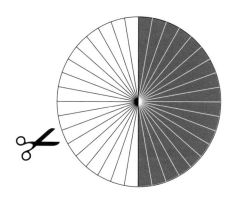

- Explain why these diagrams show that the formula for the area A of a circle of radius r is

$$A = \pi r^2 .$$

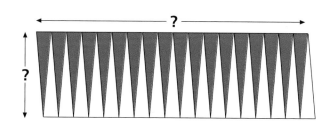

A1 Calculate, to the nearest 0.1 cm², the area of a circle whose radius is

    (a) 6 cm   (b) 7.5 cm   (c) 2.4 cm   (d) 5.6 cm   (e) 0.9 cm

A2 (a) What is the radius of this circle?

    (b) Calculate the area of the circle, to the nearest 0.1 cm².

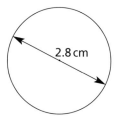

A3 Calculate, to the nearest 0.1 cm², the area of a circle with

    (a) radius 4.4 cm   (b) diameter 7.6 cm   (c) radius 8.5 cm

    (d) radius 2.7 cm   (e) diameter 9.0 cm   (f) diameter 5.3 cm

A4 Calculate the area of

    (a) the outer circle, of radius 1.6 cm

    (b) the inner circle, of radius 1.2 cm

    (c) the purple space between the two circles

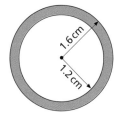

A5 Look at the diagram on the left.

    (a) Do you think the area of the light green ring is greater than, equal to, or less than the area of the green circle?

    (b) Find out whether you are right.

B Area and circumference

Give all answers to three significant figures.

B1 Calculate (i) the circumference, (ii) the area
of a circle of radius

(a) 2.2 cm   (b) 1.6 cm   (c) 5.1 cm   (d) 4.3 cm   (e) 16.2 cm

B2 (a) Calculate the area of a circle of radius 6.7 cm.

     (b) Calculate the circumference of a circle of radius 9.7 cm.

     (c) Calculate the area of a circle of **diameter** 24.3 cm.

B3 The curves at the end of a running track are semicircles.

     (a) Mary runs round the outside of the track.
In the same time John runs round the inside.
How much further does Mary run?

     (b) Calculate the area of the track (coloured).

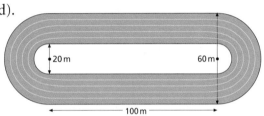

B4 A square picture is centred in a circular frame.
The diameter of the frame is 20 cm.
The corners of the picture are 5 cm
from the edge of the frame.

Calculate the area of the frame (shown purple).

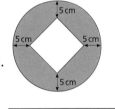

B5 Six coins all the same size are packed tightly
into a rectangular tray as shown.
The diameter of each coin is 3 cm.

     (a) How much space (shown pink) is wasted?

     (b) What percentage of the tray is wasted?

B6 Two equal semicircles overlap
What can you say about the two areas?

Explain your answer.

49

C Calculating radius given area

This is the flow diagram for the formula $A = \pi r^2$.

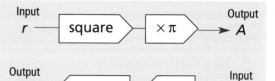

Reversing the diagram leads to the formula

$$r = \sqrt{\dfrac{A}{\pi}}$$

C1 What is the radius of a circle with area $12\,\text{cm}^2$?
Show your working.

C2 Find the radius of a circle with area

    (a) $20\,\text{cm}^2$ (b) $50\,\text{cm}^2$ (c) $120\,\text{cm}^2$
    (d) $3\,\text{cm}^2$ (e) $0.9\,\text{cm}^2$ (f) $1.5\,\text{cm}^2$

C3 The area of a circle is $35\,\text{cm}^2$.

    (a) Calculate the radius, leaving the result in the calculator display.

    (b) Use the result to calculate the circumference of the circle, to the nearest $0.1\,\text{cm}$.

    (c) Why should you **not** round the radius before using it to calculate the circumference?
         (Try doing the calculations again, but this time rounding the radius.)

C4 Calculate, to the nearest $0.1\,\text{cm}$, the circumference of a circle of area $43.5\,\text{cm}^2$.

C5 Copy this table and fill in
the missing values.

Radius	Diameter	Circumference	Area
12.8 cm			
	55.8 cm		
		72.6 cm	
			64.2 cm²

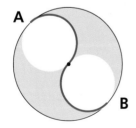

C6 The area of the large circle is $100\,\text{cm}^2$.
What is the distance AB along the red curve?

*C7 (a) Explain why the area of the square here is $2r^2$.

    (b) The blue area is $60\,\text{cm}^2$.
        Calculate the value of r.

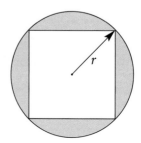

D Using exact values

The radius of the circle shown here is 3 cm.

Its circumference is $2\pi \times 3$ cm.

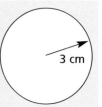

3 cm

- We can write this expression more simply as **6π** cm.
  This result is **exact**, because we have not replaced π by an approximation.

- The exact value for the area of the circle is $\pi \times 3^2 = \mathbf{9\pi}$ cm².

D1 Write down the exact value of

(a) the circumference of a circle of radius 5 cm

(b) the area of a circle of radius 5 cm

D2 Find the exact value of (i) the circumference, (ii) the area, of a circle of radius

(a) 7 cm   (b) 10 cm   (c) 15 cm   (d) 20 cm   (e) 2.5 cm

D3 What is the radius of a circle whose circumference is 9π cm?

D4 (a) Explain why the exact value, in cm², of the green area in diagram A is $36 - 9\pi$.

(b) Find an expression for the exact value, in cm², of the green area in diagram B.

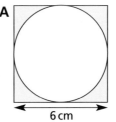

A

6 cm

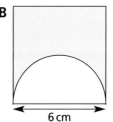

B

6 cm

D5 These designs are all made from parts of circles with radius 2 cm.
Calculate the exact value of each blue area.

(a)

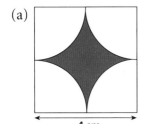

4 cm

(b)

4 cm

(c)

2 cm

D6 In both of these designs the large circle has radius 2 cm.
The white circles touch at its centre.

The small green circles touch at the centres of the other circles.

What is the exact area of the green part of each design?

(a)   (b)

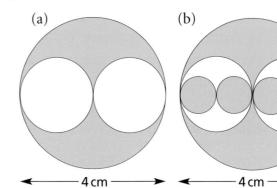

4 cm   4 cm

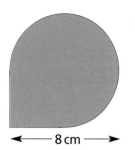

D7 This shape is made from a square and a circle with radius 4 cm. The corner of the square is at the centre of the circle.

What is the area of the shape?

← 8 cm →

D8 Find, in terms of π, an expression for the exact value of

    (a) the diameter of a circle whose circumference is 50 cm

    (b) the radius of a circle whose circumference is 20 cm

    (c) the radius of a circle whose area is 30 cm^2

E Cylinders

The volume of a cylinder can be found by multiplying the area of the base (or cross-section) by the height.

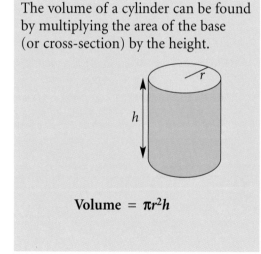

Volume = $\pi r^2 h$

The surface area is made up of two circles and a rectangle.

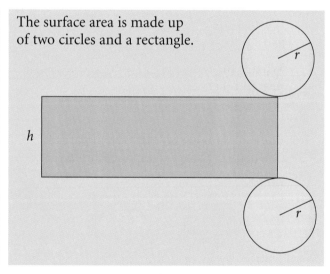

E1 The curved surface of the cylinder shown above opens out into a rectangle. One dimension of this rectangle is h.

    (a) Write down an expression for the other dimension.

    (b) Write down an expression for the area of the rectangle.

    (c) Explain why the total surface area of the cylinder can be written as $2\pi r(r + h)$.

E2 A cylinder is 7.5 cm high and has a base of radius 2.8 cm. Calculate, to 3 s. f., (a) the volume, (b) the surface area of the cylinder.

E3 Repeat E2 for a cylinder 5.8 cm high with a base radius of 6.6 cm.

E4 Copy and complete this table
of cylinders.
Give the values correct to 3 s.f.

Base radius	Base area	Height	Volume	Surface area
4 cm		9 cm		
	24 cm²	10 cm		
		6 cm	45 cm³	
3 cm			76 cm³	

E5 The **diameter** of a cylinder is 15.4 cm and the height is 4.8 cm.
Calculate, to 3 s.f., (a) the volume (b) the surface area

E6 Find an expression in terms of π for the exact value of the surface area
of a cylinder whose radius is 7 cm and height 8 cm.

*E7 Find, to 3 s.f., the volume of a cylinder whose radius is 6 cm and surface area 345 cm².

F Mixed questions

F1 Four identical white circles fit inside a coloured circle
as shown. The middle two white circles touch at the
centre of the coloured circle.

The distance along the red curve is 24 cm.

What is the radius of the coloured circle, to the
nearest 0.1 cm?

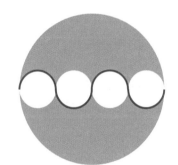

F2 Calculate, in terms of π, the exact values of

(a) the perimeter of the blue shape

(b) the perimeter of the yellow shape

(c) the area of the blue shape

(d) the area of the yellow shape

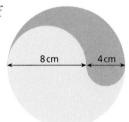

8 cm   4 cm

F3 Find, in terms of π, the exact value of

(a) the volume of a cylinder of radius 4 cm and height 5 cm

(b) the radius of a circle of circumference 30 cm

(c) the radius of a circle of area 20 cm²

(d) the height of a cylinder of radius 4 cm and volume 80 cm³

*F4 This shape consists of a square and two semicircles.
Its total area is 100 cm².

Calculate the radius of each semicircle, to the
nearest 0.1 cm.

A girdle round about the Earth

The Earth is roughly a sphere of radius 6400 km.

Imagine that an iron band is fitted tightly round the Equator.
Someone cuts the band and inserts an extra piece of length 1 metre.
As a result, the band no longer fits tightly.

If the gap is the same all round the band, how big do you think the
gap is?

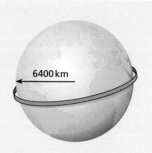

6400 km

What progress have you made?

Statement

Evidence

I can find the areas and the
circumferences of circles.

1 Calculate, to one decimal place, (i) the area,
   (ii) the circumference, of a circle of radius

   (a) 5 cm   (b) 1.7 cm   (c) 52 cm

I can find the radius of a circle if
I know the area.

2 Calculate, to the nearest 0.1 m, the radius of
   a circle with area

   (a) $9 \, m^2$ (b) $500 \, m^2$ (c) $0.1 \, m^2$

I can find the area of a circle if
I know its circumference.

3 Calculate, to the nearest $0.1 \, m^2$, the area of
   a circle with circumference

   (a) 1.5 m   (b) 38 m   (c) 300 m

I can find exact values as expressions
in terms of π.

4 Find, in terms of π, the exact value of

   (a) the circumference of a circle of radius 7.5 cm

   (b) the area of a circle of radius 1.5 cm

   (c) the diameter of a circle with
       circumference 40 cm

I can find the volume and surface area
of a cylinder.

5 Calculate, to 3 s.f., the volume and the
   surface area of a cylinder of radius 4.2 cm and
   height 6.5 cm.

Review 1

1 Calculate, to the nearest 0.1 cm,

   (a) the circumference of a circle of diameter 2.6 cm

   (b) the circumference of a circle of radius 8.8 cm

   (c) the diameter of a circle of circumference 14.6 cm

   (d) the radius of a circle of circumference 11.4 cm

2 Solve each of these equations.

   (a) $3x - 7 = 7x + 17$ (b) $11 - 2x = 3(x - 3)$ (c) $4(5 - x) = 3(11 - x)$

3 A train runs from Aysted to Creeksey and back again.
   It stops at Beaford on the way there but not on the way back.
   This graph shows its journey.

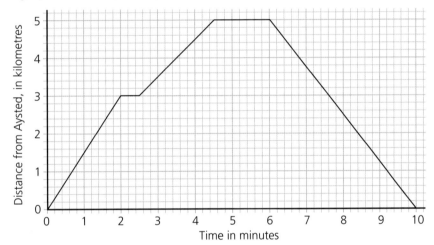

   (a) For how long does the train stop at (i) Beaford   (ii) Creeksey
   (b) What is the average speed of the train, in km/min, between

        (i) Aysted and Beaford   (ii) Beaford and Creeksey   (iii) Creeksey and Aysted

   (c) Convert each of the answers in (b) to kilometres per hour.

4 The diagram shows a journey in three parts.
   The first part is the vector from A to B.

   (a) Write AB as a column vector.

   The second part, from B to C, is the vector $\begin{bmatrix} 14 \\ -4 \end{bmatrix}$.

   The third part is from C to A.

   (b) Write the vector from C to A as a column vector.

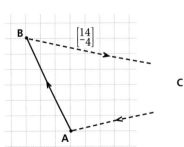

5 Multiply out the brackets from each of these expressions.

    (a) $3x(5 - 2x)$ (b) $2a(4a - b)$ (c) $pq(p + 3q)$ (d) $x^2(x - 3y)$ (e) $abc(2b + 5c)$

6 Calculate, to three significant figures,

    (a) the area of a circle of radius 7.3 cm

    (b) the area of a circle of diameter 10.4 cm

    (c) the radius of a circle of area 98.4 cm^2

7 Water flows out of a tank at a steady rate of 2.5 litres per second.
  At present there are 135 litres in the tank.
  How long will it take for the tank to empty?

8 Peter and Grant collected conkers.
  Peter collected n conkers. Grant collected three times as many as Peter.

  Grant gave Peter 20 conkers. Afterwards he had only half as many as Peter.
  Form an equation for n and solve it.

9 Write the exact value of each of the following, leaving π in your answer.

    (a) The circumference of a circle of radius 2.5 cm.

    (b) The area of a circle of radius 2.5 cm.

    (c) The volume of a cylinder of radius 2.5 cm and height 4 cm.

    (d) The surface area of a cylinder of radius 2.5 cm and height 4 cm.

10 Find the values of p, q, r and s in these vector equations.

    (a) $\begin{bmatrix} 5 \\ -1 \end{bmatrix} + \begin{bmatrix} p \\ 4 \end{bmatrix} + \begin{bmatrix} 4 \\ q \end{bmatrix} = \begin{bmatrix} -2 \\ 5 \end{bmatrix}$ (b) $\begin{bmatrix} r \\ 9 \end{bmatrix} + \begin{bmatrix} -5 \\ s \end{bmatrix} = \begin{bmatrix} 3 \\ r \end{bmatrix} + \begin{bmatrix} -1 \\ -2 \end{bmatrix}$

11 (a) Find the gradient and the intercept on the y-axis
       of each of the straight-line graphs shown here.

    (b) Write down the equation of each line.

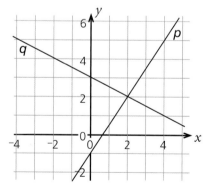

12 Multiply out the brackets from each of these expressions and simplify
  as far as possible.

    (a) $3x - 2(5 - 2x)$ (b) $2a + 3(4a - b)$ (c) $3p - 2(p + 3q)$ (d) $5x - 3(2 - 4x)$

13 Solve these equations.

    (a) $10x - (8 + 3x) = 20$ (b) $13 - 2(x - 3) = x + 1$ (c) $40 - (2x - 5) = 12$

Over to you

This is a collection of problems.
Working on them will help you

◆ think about how you might try to solve a problem

◆ try out ideas and change your approach if necessary

◆ explain your reasoning

1 Choose three consecutive numbers and add them together.
Do you always get a multiple of 3?

What happens with four consecutive numbers? five consecutive numbers?

Can you find a general rule?

Can you explain any of the rules you find?

2 Two consecutive numbers are multiplied.
Explain why the last figure of the answer cannot be 5.

What can the last figure be? Explain your answer.

3 From P to R through Q is 90 miles.
From Q to P through R is 60 miles.
From R to Q through P is 80 miles.

Calculate the direct distances PQ, QR and RP. Explain
how you worked them out.

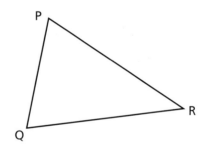

4 The diagram on the right shows how four sheets are folded to make
a newspaper with 16 pages.

The sheet with page 6 on it is shown below.
What are the numbers of the other pages on this sheet (A, B and C)?

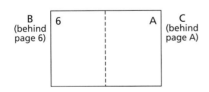

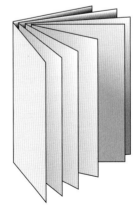

This sheet comes from a newspaper with 24 pages.
One of the pages on this sheet is page 16.
What are the numbers of the other pages on the sheet?
Copy the diagram and show where the four numbers are.

5   Draw some square 'chessboards'.
    Count the number of red squares and the number of white squares.

    Can you find a rule for working out how many of each colour there are
    when you know the size of the board?

    Can you find any rules for rectangular 'chessboards'
    (for example, 5 by 3)?

6   (a) Jenny's calculator has a peculiar fault.
        When she presses the 5 key, the calculator treats it as 8.
        When she presses 8, the calculator treats it as 5.
        The same thing happens in the display: 8 means 5 and 5 means 8.

        What does Jenny's calculator show when she does 457 + 328?

    (b) Ruth's calculator is like Jenny's but instead of 5 and 8, a different pair of
        digits is interchanged.

        When she does 712 + 453 she gets 369. Which digits are interchanged?
        Explain how you got your answer.

7   Calculate the value of x.
    Explain how you worked it out.

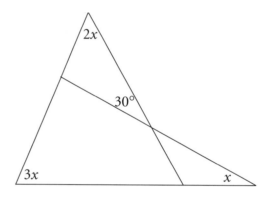

8   Calculate the angle between the hands of a clock at 3:40 p.m.
    Explain how you worked it out.

9   This fruit is to be split between three   13 apples   18 bananas   20 oranges
    people: Ann, Bob and Ceri.

    Each person gets the same number of fruits in all, but
        Ann doesn't want bananas, and wants more apples than oranges;
        Bob wants all three, but more bananas than apples;
        Ceri doesn't want oranges, but wants more bananas than apples.
    How can you split up the fruit? Are there different ways to do it?

Trial and improvement

This will help you solve problems and algebraic equations using trial and improvement methods.

A Introducing the method

Problem 1

I think of a number. I add 7. I multiply the result by my starting number.
The answer is 408.

- What number did I start with?

Problem 2

The length of this rectangle is 2 cm more than its width.

- Find the width, x cm, if the area is
  (a) 35 cm²   (b) 48 cm²   (c) 40 cm²

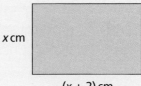

x cm

$(x + 2)$ cm

Problem 3

A farmer is cutting the wheat in a square field.
She harvests the wheat by cutting inwards from the edges of the field.

By the time she has cut all the wheat within a 20 metre band inside the edge of the field, half the harvest is complete.

- What size is the field?

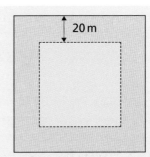

20 m

Problem 4

A toolmaker needs to make a steel cuboid.
The width must be half the length, and the height must be half the width.
The volume of the cuboid must be 100 cm³.

- What are the dimensions of the cuboid?

Contest between two people

Who needs fewer trials to solve this problem?

12768 tiles, each 1 cm square, can be arranged to make a rectangle whose length is 2 cm more than its width.

What are the dimensions of the rectangle?

B Solving equations

Problem

Make me a cuboid with a square base please. Volume must be 7 cm³ and the height 2 cm more than the base. Ta - Jen

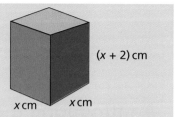

$(x + 2)$ cm

x cm x cm

Let the side of the base be x cm. The height is then $(x + 2)$ cm.
The volume of the cuboid, in cm³, is $x \times x \times (x + 2) = x^2(x + 2) = x^3 + 2x^2$

So we have to solve the equation

$$x^3 + 2x^2 = 7$$

- We need two **starting values** for x, one for which $x^3 + 2x^2$ is less than 7 and one for which $x^3 + 2x^2$ is greater than 7.

  The solution will be somewhere between these two starting values.

  When $x = 1$, the value of $x^3 + 2x^2$ is 3.
  When $x = 2$, the value of $x^3 + 2x^2$ is 16.

  So the solution must be between 1 and 2.

- We can show this in a table …   … or on a number line.

x	$x^3 + 2x^2$	
1	3	result too small
2	16	result too big

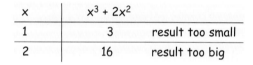

- Try $x = 1.5$.

x	$x^3 + 2x^2$	
1	3	result too small
2	16	result too big
1.5	7.875	result too big

- The result for $x = 1.5$ was quite close, so try $x = 1.4$.

x	$x^3 + 2x^2$	
1	3	result too small
2	16	result too big
1.5	7.875	result too big
1.4	6.664	result too small

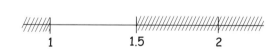

- Now try values of x between 1.4 and 1.5, such as 1.43, and so on.

- After working with values of x to two decimal places, you will find that the solution lies between 1.42 and 1.43.

x	$x^3 + 2x^2$	
1.42	6.896088	result too small
1.43	7.014007	result too big

- To decide whether the solution is closer to 1.42 or 1.43, try **1.425** (halfway between).

x	$x^3 + 2x^2$	
1.42	6.896088	result too small
1.43	7.014007	result too big
1.425	6.954890625	result too small

The solution lies between 1.425 and 1.43, so it is closer to 1.43 than to 1.42.

The solution is $x = 1.43$ **correct to two decimal places**.

(So the dimensions of the cuboid are 1.43 cm, 1.43 cm and 3.43 cm.)

B1 The equation $x^2 + x = 10$ has a solution somewhere between $x = 2$ and $x = 3$.
Use trial and improvement to find the solution correct to two decimal places.

B2 The equation $x^3 - x^2 = 3$ has a solution somewhere between $x = 1$ and $x = 2$.
Use trial and improvement to find the solution correct to two decimal places.

B3 The height of a square-based cuboid is 1 cm less than the side of the base.
The volume of the cuboid is 5 cm³.
Find the dimensions of the cuboid, correct to two decimal places.

B4 (a) What is the value of $8x - x^3$ when $x = 2$?
(b) What is the value of $8x - x^3$ when $x = 3$?
(c) Find, correct to 2 d.p., the value of x between 2 and 3 for which $8x - x^3 = 1$.

B5 (a) There is a value of z, greater than 1, that makes $3z - z^2 = 1.5$.
Find this value of z correct to two decimal places.
(b) There is another value of z that makes $3z - z^2 = 1.5$.
Find this value correct to two decimal places.

B6 The area of this triangle is 10 cm².
What is its height, correct to 3 d.p.?

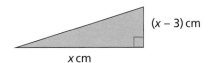

$(x - 3)$ cm

x cm

*B7 Find, correct to two decimal places, the dimensions of a rectangle whose perimeter is 20 cm and whose area is 20 cm².

C Using a spreadsheet

With a spreadsheet it is quicker to enter a formula to work out intermediate values, rather than to choose some yourself.

For example, the equation $x^3 + 4x^2 - 7 = 0$ has a solution between 1 and 2.
A formula can be used to try $x = 1.1$, $x = 1.2$, $x = 1.3$, ...

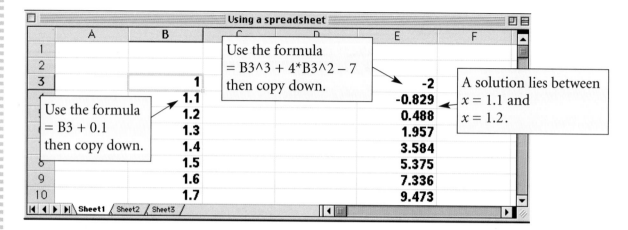

To find the solution between 1.1 and 1.2 more accurately, you can insert 9 rows between 1.1 and 1.2 and use them for $x = 1.11, 1.12, 1.12, ...$

C1 (a) Create the spreadsheet above and use it to find the solution of the equation between 1 and 2 accurate to 4 d.p.

   (b) There are two other values of x that make $x^3 + 4x^2 - 7 = 0$.
   Find these other two solutions accurate to 4 d.p.

C2 The equation $x^5 - 2x^2 + 1 = 0$ has three solutions between $^-2$ and 2.
   Use a spreadsheet to find each of the solutions to four decimal places.

C3 Use a spreadsheet to solve this problem.

   A rectangle's length is 2 cm more than its width.
   Its area is 4 cm². Calculate the width, as accurately as you can.

What progress have you made?

Statement	Evidence
I can use trial and improvement to solve algebraic equations.	1 The equation $x^3 - 4x = 2$ has a solution between 2 and 3. Use trial and improvement to find this solution to two decimal places.

⑨ Exploring Pascal's triangle

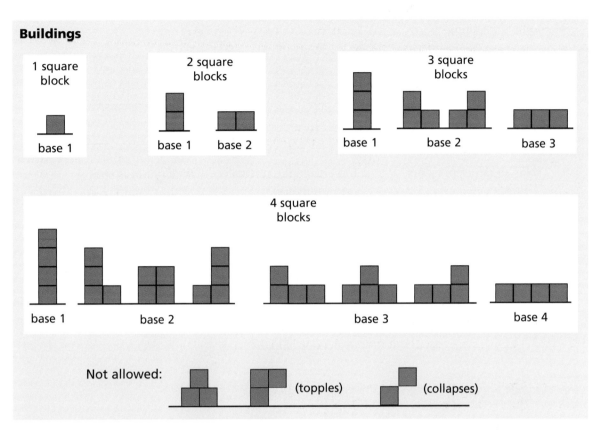

Buildings

1 square block

base 1

2 square blocks

base 1   base 2

3 square blocks

base 1   base 2   base 3

4 square blocks

base 1   base 2   base 3   base 4

Not allowed: (topples)   (collapses)

Two-colour towers

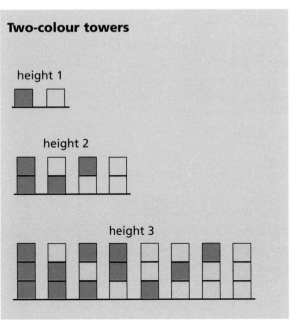

height 1

height 2

height 3

Routes on a grid

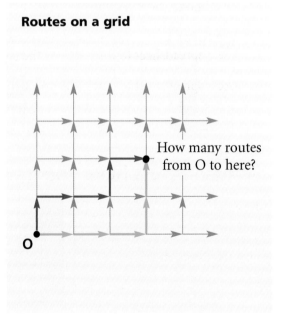

How many routes from O to here?

O

⑩ Calculating probabilities

This work will help you

◆ multiply fractions

◆ understand independent events

◆ calculate probabilities

A Fractions of fractions

This square represents 1 unit.

First it is split vertically into **quarters**.

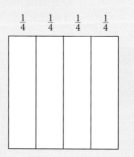

Then it is split horizontally into **thirds**.

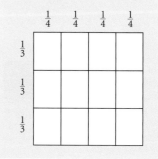

Lightly shade $\frac{1}{4}$.

Then heavily shade $\frac{1}{3}$ of $\frac{1}{4}$.

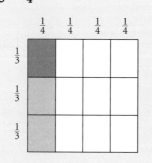

What fraction of the square is $\frac{1}{3}$ of $\frac{1}{4}$?

• What does each of these diagrams show?

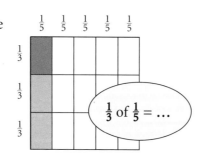

$\frac{1}{3}$ of $\frac{1}{5}$ = ...

... of ... = ...

A1 What does each of these diagrams show?

(a)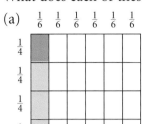

(b)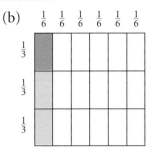

(c)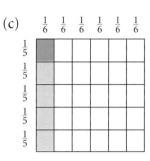

A2 I lightly shade $\frac{3}{4}$,

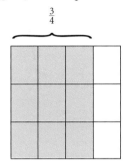

and heavily shade $\frac{2}{3}$ of $\frac{3}{4}$.

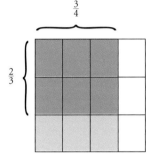

What fraction of the square is $\frac{2}{3}$ of $\frac{3}{4}$?

A3 I lightly shade $\frac{3}{5}$,

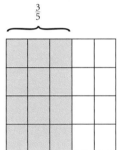

and heavily shade $\frac{3}{4}$ of $\frac{3}{5}$.

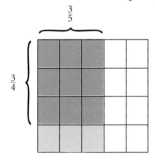

What fraction of the square is $\frac{3}{4}$ of $\frac{3}{5}$?

A4 Draw this diagram.

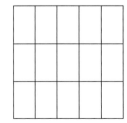

By light and dark shading, show some different fractions of fractions (at least three!).

Here is one to start you off.

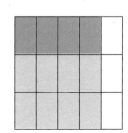

$\frac{1}{3}$ of $\frac{4}{5}$ = ...

A5 Draw a diagram to show that $\frac{2}{5}$ of $\frac{2}{3}$ = $\frac{2}{3}$ of $\frac{2}{5}$.

65

B Multiplying fractions

$\frac{1}{2}$ of 8 and $\frac{1}{2} \times 8$ are equal. (They are both 4.)

In a similar way we write $\frac{1}{2}$ of $\frac{1}{4}$ as $\frac{1}{2} \times \frac{1}{4}$.

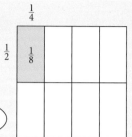

$$\frac{1}{2} \times \frac{1}{4} = \frac{1}{8}$$

B1 Work these out. (You can use a square diagram if you like.)

(a) $\frac{1}{2} \times \frac{1}{3}$ (b) $\frac{1}{3} \times \frac{1}{4}$ (c) $\frac{1}{3} \times \frac{1}{6}$ (d) $\frac{1}{4}$ of $\frac{1}{3}$ (e) $\frac{1}{2} \times \frac{1}{5}$

B2 Work these out.

(a) $\frac{1}{2} \times \frac{3}{5}$ (b) $\frac{3}{5} \times \frac{3}{4}$

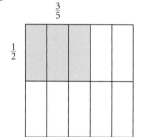
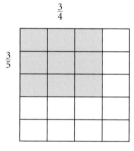

(c) $\frac{3}{4} \times \frac{1}{5}$ (d) $\frac{2}{3}$ of $\frac{4}{5}$ (e) $\frac{3}{5} \times \frac{3}{8}$ (f) $\frac{2}{5}$ of $\frac{2}{5}$ (g) $\frac{2}{5} \times \frac{2}{3}$

B3 Work these out.

(a) $\frac{2}{3} \times \frac{3}{4}$ (b) $\frac{2}{5}$ of $\frac{3}{8}$ (c) $\frac{3}{4} \times \frac{2}{5}$ (d) $\frac{2}{3}$ of $\frac{5}{6}$ (e) $\frac{5}{6} \times \frac{3}{4}$

B4 How can you work out $\frac{2}{3} \times \frac{4}{5}$ without drawing a diagram?

B5 Work these out without drawing diagrams.

Simplify the result where you can.
(For example, you could simplify $\frac{6}{8}$ to $\frac{3}{4}$.)

(a) $\frac{2}{3} \times \frac{3}{8}$ (b) $\frac{3}{4} \times \frac{3}{8}$ (c) $\frac{2}{3}$ of $\frac{2}{5}$ (d) $\frac{3}{4} \times \frac{5}{8}$ (e) $\frac{5}{6} \times \frac{3}{5}$

C Traffic flows

Diagrams like this are used to show how the traffic out of a
town goes on to different routes.

When the traffic gets to A, it splits up.
$\frac{1}{3}$ goes to B and $\frac{2}{3}$ goes to C.

The traffic arriving at B splits
$\frac{1}{2}$ to D and $\frac{1}{2}$ to E.

So the traffic arriving at D
is $\frac{1}{2}$ of $\frac{1}{3}$ of the total,

which is $\frac{1}{6}$ of the total.

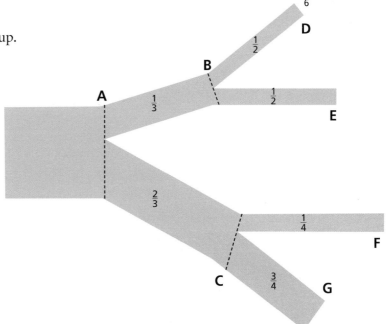

C1 What fraction of the total traffic arrives at

    (a) E   (b) F   (c) G

    Check that the fractions arriving at D, E, F, G add up to 1.

C2 This diagram is similar to the one above,
but simplified.

    What fraction of the traffic leaving O
arrives at

    (a) R   (b) S   (c) T   (d) U

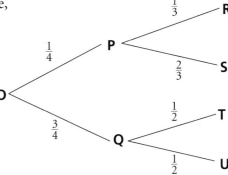

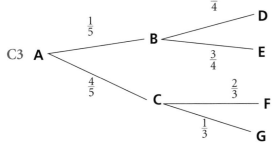

C3 A

(a) In this diagram, what fraction of the
traffic leaving A arrives at

    (i) D   (ii) E   (iii) F   (iv) G

(b) If 60 cars leave A, how many of them
would you expect to arrive at each of
D, E, F and G?

D Independent events

Here are a bag and a spinner.

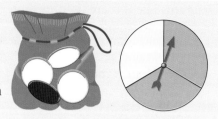

Suppose you take out, at random, a counter from the bag and you spin the arrow.
Taking a counter from the bag will not influence what colour the arrow points to. We say that the outcomes from the bag and the spinner are **independent**.

This diagram shows that there are 12 equally likely pairs of outcomes.

Spinner

G

G

Y

B W W W

Counter

Suppose you want to know the probability that the counter is black and the arrow points to yellow. The shaded part represents the only way this can happen.

The probability of 'black, yellow' is $\frac{1}{12}$.

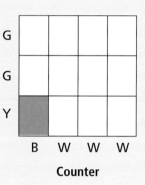

Spinner G

G

Y

B W W W

Counter

We can get this result by multiplying probabilities.

probability of black counter	×	probability of yellow on spinner	=	probability of getting black counter **and** yellow on spinner
$\frac{1}{4}$	×	$\frac{1}{3}$	=	$\frac{1}{12}$

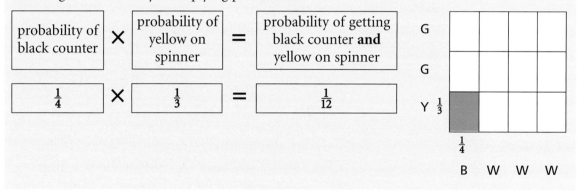

68

D1 For the bag and spinner on the opposite page, work out the probability that

(a) the counter is black and the arrow points to green

(b) the counter is white and the arrow points to green

(c) the counter is white and the arrow points to yellow

D2 Sarah spins the arrows on these two spinners.
Work out the probability that both arrows
point to red.

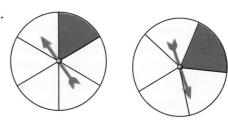

D3 The square diagram shows all the equally likely
outcomes when a counter is taken at random from
each of these two bags.

The shaded part shows all the outcomes where
both counters are black.

(a) What is the probability that the counter taken
from bag P is black?

(b) What is the probability that the counter taken
from bag Q is black?

(c) What is the probability that both counters
are black?

(d) What do you do to (a) and (b) to work out (c)?

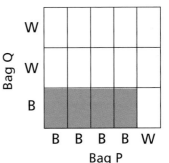

D4 A counter is taken at random from each
of these two bags.

Work out the probability that both
counters are black.

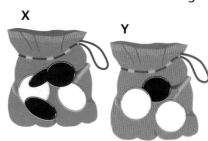

D5 These two spinners are spun.
Work out the probability that both arrows
point to red.

E Tree diagrams

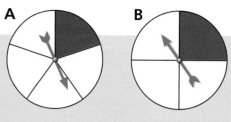

This **tree diagram** shows the outcomes when each arrow is spun.

We write the probability of each separate outcome on its branch.

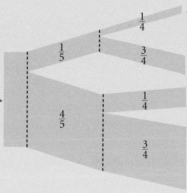

The heavy lines show the outcome

   A red, B red

whose probability is $\frac{1}{5} \times \frac{1}{4} = \frac{1}{20}$.

You can think of it as a 'traffic flow' like this.

Imagine that both arrows are spun a large number of times.

'Both red' occurs $\frac{1}{20}$ of the times.

E1 Calculate the probability of each of these outcomes.

   (a) A red, B white   (b) A white, B red   (c) A and B both white

E2 Copy and complete the tree diagram for this pair of spinners.

Calculate the probability of each of these outcomes.

   (a) P and Q both red

   (b) P and Q both white

   (c) P red and Q white

   (d) P white and Q red

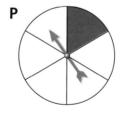
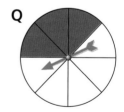

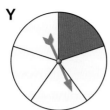
X   Y

E3 (a) Draw a tree diagram for this pair of spinners.

(b) Calculate the probability of each of the four possible outcomes when both arrows are spun.

E4 A coin is spun and a dice is thrown.

(a) Copy and complete this tree diagram.

(b) Calculate the probability of getting

    (i) head and six   (ii) head but not six

    (iii) tail and six   (iv) tail but not six

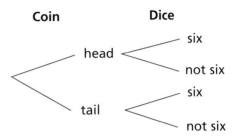

Look again at the tree diagram at the start of this section.

The four different outcomes of the two spins are

'A red, B red' 'A red, B white' 'A white, B red' 'A white, B white'

Now think about the event 'the two spinners give the same colour'.
This can happen in two different ways: 'A red, B red' or 'A white, B white'.

The probability of 'A red, B red' is $\frac{1}{5} \times \frac{1}{4} = \frac{1}{20}$.

The probability of 'A white, B white' is $\frac{4}{5} \times \frac{3}{4} = \frac{12}{20}$.

The probability that the two spinners give the same colour is $\frac{1}{20} + \frac{12}{20} = \frac{13}{20}$.

E5 Calculate the probability that the two spinners in question E2 give the same colour when spun.

E6 Do the same for the two spinners in question E3.

E7 Calculate the probability that the two spinners in question E2 give **different** colours when spun.

E8 Do the same for the two spinners in question E3.

E9 A counter is taken at random from each of these bags. Calculate the probability that the two counters are

(a) the same colour   (b) different colours

E10 Do the same for these bags.

71

F Adding fractions: a reminder

It is easy to add fractions with the same denominator: $\frac{3}{20} + \frac{10}{20} = \frac{13}{20}$

When the fractions have different denominators, you need to use **equivalent fractions**.

Example $\frac{3}{8} + \frac{1}{6}$

For the denominator, choose a number which is a multiple of both 8 and 6.
24 is the smallest.

$\frac{3}{8} = \frac{9}{24}$ and $\frac{1}{6} = \frac{4}{24}$

So $\frac{3}{8} + \frac{1}{6} = \frac{9}{24} + \frac{4}{24} = \frac{13}{24}$.

F1 Work these out.

(a) $\frac{1}{5} + \frac{1}{3}$ (b) $\frac{1}{4} + \frac{1}{3}$ (c) $\frac{2}{5} + \frac{1}{3}$ (d) $\frac{3}{8} + \frac{2}{5}$ (e) $\frac{3}{10} + \frac{1}{6}$

F2 A spinner has three sections, red, blue and green.
The probability of getting red is $\frac{1}{4}$ and of blue $\frac{1}{5}$.

Calculate the probability of getting (a) either red or blue   (b) green

F3 A spinner has three sections, red, green and yellow.
The probability of getting red is $\frac{2}{3}$ and of green $\frac{1}{8}$.

Calculate the probability of getting (a) either red or green   (b) yellow

What progress have you made?

Statement	Evidence
I can multiply fractions.	1 Work these out. (a) $\frac{3}{4} \times \frac{1}{5}$ (b) $\frac{3}{8} \times \frac{4}{5}$
I can work out the probability that two independent events happen.	2 A counter is picked at random from each of these bags. 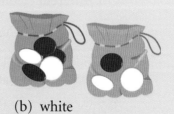 What is the probability that both counters are (a) black (b) white
I can draw a tree diagram and calculate probabilities from it.	3 Draw a tree diagram for the possible outcomes of picking a counter from each bag above. Calculate the probability of getting one counter of each colour.

⑪ Linear equations and graphs

This work will help you

◆ draw graphs from algebraic equations

◆ solve simultaneous equations graphically

A Equations, tables, graphs

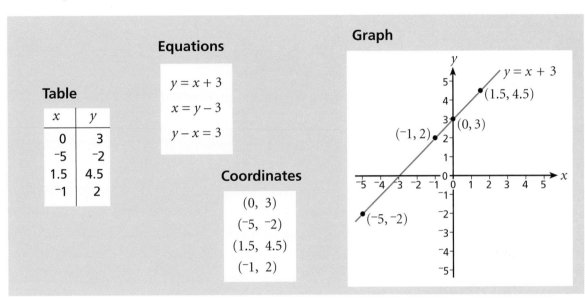

Table

x	y
0	3
−5	−2
1.5	4.5
−1	2

Equations

$y = x + 3$

$x = y − 3$

$y − x = 3$

Coordinates

(0, 3)

(−5, −2)

(1.5, 4.5)

(−1, 2)

Graph

A1 Here is a table of values of x and y.

Which of the equations below fit the values in the table?
(There is more than one that fits.)

A $y = x$

B $y = 2x − 1$

C $y + 1 = 2x$

D $y = 3x − 2$

E $2x − y = 1$

F $x = \dfrac{y + 1}{2}$

x	y
1	1
2	3
3	5
4	7

A2 Look at this table of values of x and y.
Write down equations that connect x and y

(a) with $y = \ldots$ on the left-hand side

(b) with $x = \ldots$ on the left

(c) with both x and y on the left

x	y
1	3
2	4
3	5
4	6

73

A3 For each of these tables, find some different equations that fit.

(a)

x	y
0	6
1	5
2	4
3	3

(b)

x	y
0	3
1	6
2	9
3	12

(c)

x	y
1	-2
2	-1
3	0
4	1

(d)

x	y
-5	5
0	0
1	-1
3	-3

A4 For each of these tables, find two different equations that connect a and b.

(a)

a	b
0	0
1	2
2	4
3	6

(b)

a	b
0	0
1	-3
2	-6
3	-9

(c)

a	b
0	0
2	1
4	2
6	3

(d)

a	b
1	0
3	1
5	2
7	3

A5 This table is for the equation $y = 3x - 1$.

Copy the table and fill in the gaps.

x	y
0	-1
1	2
3	...
...	17

A6 Copy these tables and their equations.
Fill in the gaps in the tables.

(a) $x = 2y - 4$

x	y
...	0
...	$\frac{1}{2}$
...	1
0	...

(b) $g + h = 10$

g	h
0	...
3	...
...	1
...	0

(c) $3s + 4t = 12$

s	t
0	...
-4	...
...	0
...	-3

(d) $y = 6 - x$

x	y
4	...
...	1
8	...
...	-6

A7 Copy and complete each of these tables.
Draw the graph of each equation, with x and y going from -4 to 4.

(a) $x = 1$

x	y
1	0
1	3
...	4
...	-3

(b) $y = 2$

x	y
0	2
-3	...
1	...
4	...

(c) $x + y = 0$

x	y
-4	4
2	...
...	1
0	...

B Different forms for the equation of a line

There are two common ways of giving the equation of a straight line.

You can give y **explicitly** as a function of x or you can give an equation in **implicit** form.

| $y = 2x + 5$ | $y = 3x - 2$ | $y = 10 - 2x$ | | $x + y = 8$ | $x + 2y = 10$ | $2x - 5y = 20$ |

If an explicit form is given, we can find the **gradient** and the **intercept** on the y-axis.

Example $y = 2x + 3$

gradient = 2

intercept = 3

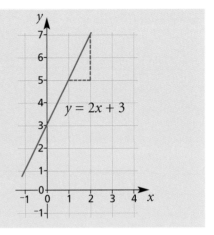

If an implicit form is given, we can easily find where the graph crosses each axis.

Example $3x + 2y = 12$

- Put $x = 0$ in the equation: $0 + 2y = 12$
$$y = 6$$

So the line crosses the y-axis at $(0, 6)$.

- Put $y = 0$ in the equation: $3x + 0 = 12$
$$x = 4$$

So the line crosses the x-axis at $(4, 0)$.

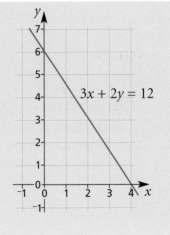

B1 Draw axes on squared paper with x and y from $^-6$ to 6.

Draw the graph of each of these equations.

(a) $y = x + 2$ (b) $y = ^-2x + 5$ (c) $x + 3y = 6$ (d) $2x - 3y = 12$

(e) $y = 2x - 4$ (f) $x - 2y = 4$ (g) $4x - 5y = ^-20$ (h) $x + 3y = 1$

*B2 (a) The line whose equation is $ax + by = 20$ goes through the points $(5, 0)$ and $(0, 4)$.
Find the values of a and b.

(b) Find the equation of the line that goes through

(i) $(7, 0)$ and $(0, 3)$ (ii) $(4, 0)$ and $(0, ^-2)$

C Parallel lines

The line $y = 2x + 3$ has gradient 2.
So do the lines

$$y = 2x + 1, \quad y = 2x - 3, \quad y = 2x, \quad \text{and so on.}$$

It follows that these are all parallel lines.
Any other line parallel to these will have
an equation of the form

$$y = 2x + \text{something.}$$

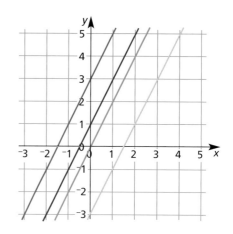

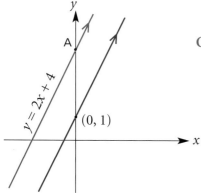

C1 (a) The line $y = 2x + 4$ crosses the y-axis at A.
What are the coordinates of A?

   (b) A second line parallel to $y = 2x + 4$
goes through the point $(0, 1)$.
What is the equation of this line?

C2 (a) The line $y = -\frac{1}{2}x + 3$ crosses the y-axis at K.
What are the coordinates of K?

   (b) A second line parallel to $y = -\frac{1}{2}x + 3$
goes through the point $(0, -1)$.
What is the equation of this line?

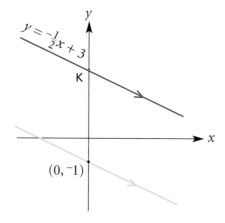

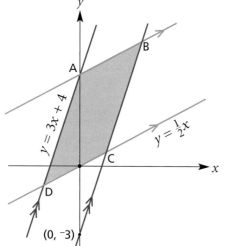

C3 In this diagram the shape ABCD is a parallelogram.

   (a) What are the coordinates of A?

   (b) What is the equation of the line CB?

   (c) What is the equation of the line AB?

C4 (a) From the coordinates of A and B, work out the gradient of the line through A and B.

(b) Write down the equation of the line through A and B.

(c) Write down the equation of the line through $(0, {}^-3)$ which is parallel to AB.

(d) Find the coordinates of the points where each of the parallel lines crosses the line $x = 5$.

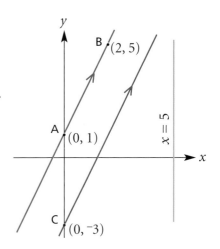

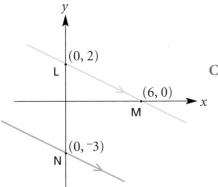

C5 (a) Find the gradient of the line through L and M.

(b) Write down the equation of the line through L and M.

(c) Write down the equation of the line through N which is parallel to LM.

(d) Find the coordinates of the point where the second line crosses the x-axis.

*C6 (a) Find the equation of the line through A and B.

(b) Imagine that the line through A and B is reflected in the y-axis.
Find the equation of the image.

(c) Now imagine that the line through A and B is reflected in the x-axis.
Find the equation of the image.

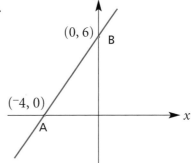

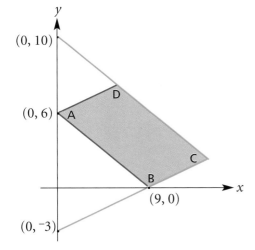

*C7 ABCD is a parallelogram.
Find the equation of

(a) AB

(b) BC

(c) CD

(e) AD

77

D Simultaneous equations

$y = x - 1$

x	y
-2	-3
-1	-2
0	-1
1	0
2	1
3	2

$x + 2y = 4$

x	y
-2	3
-1	$2\frac{1}{2}$
0	2
1	$1\frac{1}{2}$
2	1
3	$\frac{1}{2}$

$y = x - 1$

x	y
-2	-3
-1	-2
0	-1
1	0
2	1
3	2

$x + 2y = 4$

x	y
-2	3
-1	$2\frac{1}{2}$
0	2
1	$1\frac{1}{2}$
2	1
3	$\frac{1}{2}$

D1 (a) Find three points which fit the equation $y = 2x - 1$.

On squared paper, draw axes each going from about ⁻5 to 5.
Then draw and label the graph of $y = 2x - 1$.

(b) On the same axes, draw and label the line $y - x = 1$.

(c) Use your graphs to find a pair of coordinates that fits
both $y = 2x - 1$ and $y - x = 1$.
Check that your coordinates fit both equations by substituting them into both.

D2 Draw graphs on squared paper to find a point that fits both $x + y = 6$ and $y = \frac{1}{2}x$.

D3 (a) On squared paper, draw the graphs of $2x + 3y = 6$ and $y = 2x - 2$.

(b) Find the coordinates of the point that fits both equations.
Check by substituting into the equations.

D4 The lines $y = {}^{-}2$, $x + y = 6$ and $y = 2x$ form a triangle.

   (a) On squared paper, draw axes with x and y going from $^-3$ to 8. Draw and label the three lines on your axes.

   (b) What are the coordinates of the three vertices of the triangle formed by the lines?

   (c) Work out the area of the triangle.

D5 Use the graphs shown here to find, as accurately as you can, the values of x and y that satisfy each pair of simultaneous equations below.

In each case substitute the values into both equations and check that they are approximately correct.

   (a) $y = x + 1$
      $3x + 2y = 6$

   (b) $2x + y = 1$
      $y = x + 1$

   (c) $3x + 2y = 6$
      $y = \frac{1}{2}x - 2$

   (d) $2x + y = 1$
      $y = \frac{1}{2}x - 2$

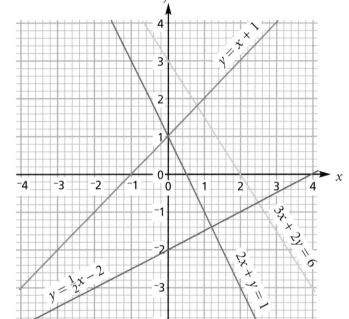

D6 (a) Draw axes on graph paper with x and y from $^-5$ to 5. Draw the lines whose equations are

        $3x + 4y = 12$
        $2x + 5y = 10$

   (b) Find, as accurately as you can, the values of x and y that satisfy both equations simultaneously.

D7 Repeat D6 for the equations

        $y = 2x - 1$
        $2x - 5y = {}^-10$

D8 Something peculiar happens when you try to solve each pair of equations below by drawing graphs.

Find out what happens in each case and try to explain it.

(a) $2x + 3y = 12$
$4x + 6y = 18$

(b) $4x + 2y = 10$
$2x + y = 5$

What progress have you made?

Statement

Evidence

I can draw the graph of a linear equation, given in either explicit or implicit form.

1 Draw axes with x and y from $^-5$ to 5. Draw the graphs of

(a) $y = 2x - 4$

(b) $3x + 5y = 15$

(c) $x - 2y = 4$

I understand the connection between the equations of lines that are parallel.

2

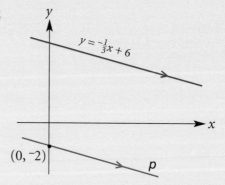

(a) What is the equation of line p?

(b) What are the coordinates of the point where p crosses the x-axis?

I can solve simultaneous equations using graphs.

3 Draw axes with x and y from $^-5$ to 5. By drawing graphs, find approximately the values of x and y that fit both of the equations

$y = 2x + 3$

$5x + 4y = 20$

12 Using and misusing statistics

This work will help you criticise the way in which statistical information is gathered or presented.

A Misleading charts and pictures

- What could be misleading about each of these charts?

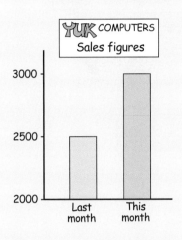

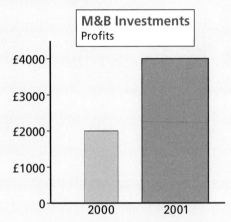

- Roughly what percentage of the pie do you think is meat, gravy or crust?

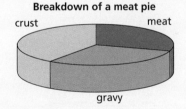

- This advertisement about the National Lottery appeared in newspapers.

  Why might it be described as misleading?

  How would you improve it?

These questions can be tackled by pairs working together.

A1 The next two charts could mislead in some way.

Find what could be misleading about each chart
and suggest how you might make it less misleading.

(a) **Workers with at least a minimum recognised qualification**

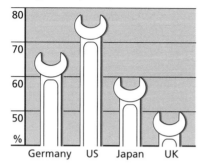

(b) **Percentage of 3–5 year olds receiving some childcare in Europe**

France	95%
Belgium	95%
Italy	88%
Denmark	87%
Spain	66%
Greece	62%
West Germany	60%
Ireland	52%
Netherlands	50%
Luxembourg	48%
United Kingdom	44%
Portugal	25%

A2 Find what could be misleading about each of these.

(a)

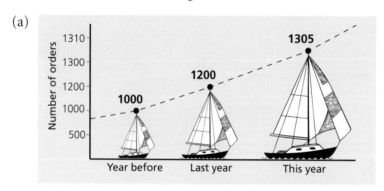

(b)

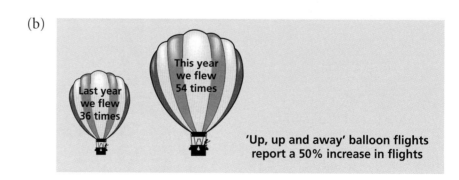

82

A3 These 3-D pie charts all show the same data about how Dwayne spends his day.
Do they all give the same impression?
How easy is it to estimate the percentage of time Dwayne spends
on each activity (or inactivity!)?

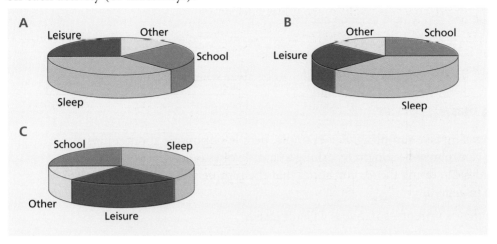

A4 Look at the two graphs here.
Which of the two businesses is growing faster?

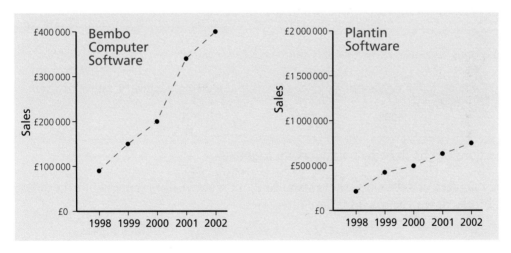

A5 These charts come from a phone company's report to customers.
Find what could be misleading about each one.

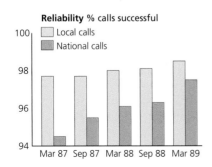

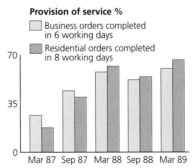

B Problems arising in collecting data

Defining categories

A magazine article claims that only 35% of the students in a college are studying a science subject.

- Why is this statement unclear as it stands?
- What do you need to know in order to be clear what it means?

Avoiding bias

Most surveys involve **sampling**. For example, people's opinions about a new shopping centre may be sought by asking a sample of people.
The organisation doing the survey hopes that the sample will be **representative** of people in general.

A sample that is unrepresentative is called **biased**.

- How are these ways of sampling shoppers likely to be biased?

Asking people as they leave the shopping centre car park

Asking people using the centre between 10 a.m. and 11 a.m. one day

Leaving a questionnaire form by each shop exit for a week

Giving a phone number in local papers and inviting people to call with their views

These questions can be done by pairs working together.

B1 The manager of a theme park wants to find out what visitors think about the rides. He breaks down visitors into these categories.

| young children | teenagers | young adults | older adults |

What is wrong with these categories as they stand?
How could you improve them?

B2 A student is going to carry out a survey of TV programmes.
She breaks them down into the following categories.

| films | dramas | soaps | comedy | factual | news and current affairs | sport |

What difficulties do you foresee with these categories?

B3 An article says '93% of all music CDs bought by young people are pop music.'
What is unclear about this statement?

B4 A magazine wants to find out what proportion of people spend their summer holiday abroad.

Criticise each of these ways of getting a representative sample of people to ask.

(a) Asking people who visit travel agents

(b) Getting a radio station to ask people to phone in and tell them

(c) Asking people as they leave banks

(d) Asking people in a town centre on a rainy day

What progress have you made?

Statement

Evidence

I can identify misleading features in graphs, charts and diagrams.

1 What could be misleading about this graph?

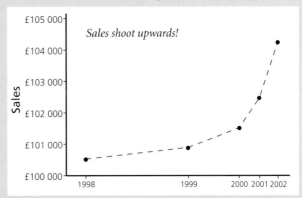

I understand about bias.

2 A magazine wants to survey people's reading habits.

People leaving a library during the day are asked their views.

In what ways would this sample of people be biased?

⓭ Pythagoras's theorem

This is about the lengths of sides of right-angled triangles.
The work will help you

♦ find the length of one side of a right-angled triangle
   if you know the lengths of the other two sides

♦ solve problems involving the lengths of sides of right-angled triangles

A Tilted squares

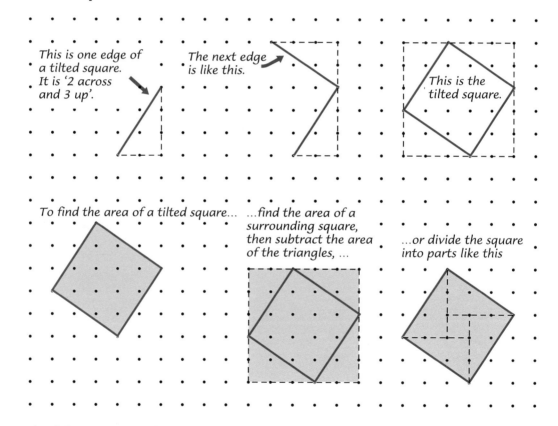

This is one edge of a tilted square. It is '2 across and 3 up'.

The next edge is like this.

This is the tilted square.

To find the area of a tilted square... *...find the area of a surrounding square, then subtract the area of the triangles, ...*

...or divide the square into parts like this

A1 Each of these is a side of a tilted square.
Draw each square and work out its area.

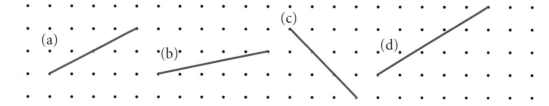

(a)   (b)   (c)   (d)

A2 Investigate the areas of the squares in this pattern.

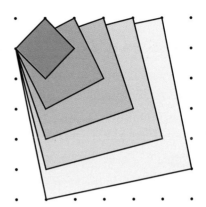

B Squares on right-angled triangles

The three squares Q, R and S are drawn on the sides of a right angled triangle

Copy the drawing on to dotty paper.
Find and record the area of each square.

Repeat this process for different right-angled triangles.

What happens?

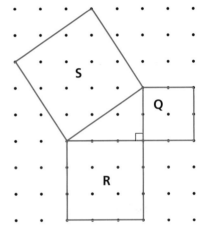

B1 Find the missing areas of the squares on these right-angled triangles.

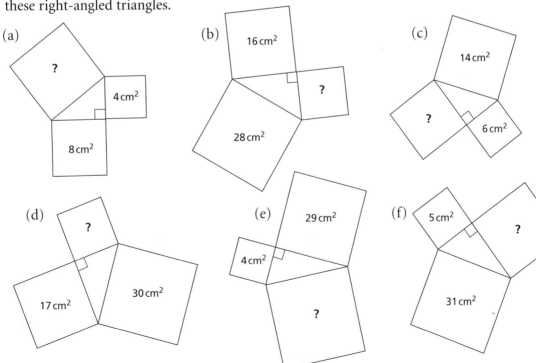

(a) ? 4 cm² 8 cm²

(b) 16 cm² ? 28 cm²

(c) 14 cm² ? 6 cm²

(d) ? 17 cm² 30 cm²

(e) 29 cm² 4 cm² ?

(f) 5 cm² ? 31 cm²

Pythagoras's theorem

In a right-angled triangle the side opposite the right angle
is called the **hypotenuse**. It is the longest side.

You have found that the area of the square on the hypotenuse
equals the total of the areas of the squares on the other two sides.

Here, area C = area A + area B

This is known as Pythagoras's theorem.
Pythagoras was a Greek mathematician and mystic.
A theorem is a statement that can be proved true.

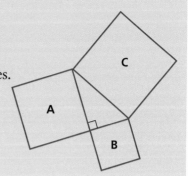

Using Pythagoras's theorem you can work with the
lengths of sides as well as the areas of squares on them.

B2 (a) What would be the area of
a square drawn on side XY?

(b) What is the length of side XY?

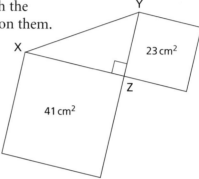

B3 What is the area of
the square drawn here? ➡

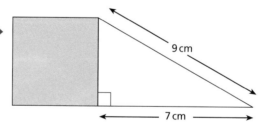

B4 Work out the missing area or length in each of these.

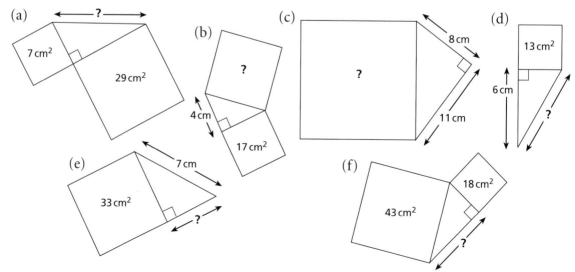

88

B5 Work out the missing area or length in each of these.

(a)　(b)　(c)　(d)

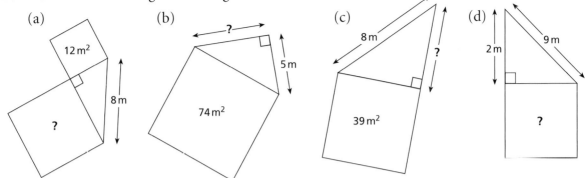

Pythagoras in practice

Pythagoras is useful for working out lengths when designing and constructing things.

You don't have to draw squares on the sides of the right-angled triangle you are using.
You can think of Pythagoras just in terms of the lengths of the sides, as shown here.

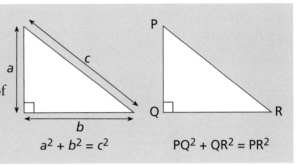

$$a^2 + b^2 = c^2$$

$$PQ^2 + QR^2 = PR^2$$

B6 (a) Use Pythagoras to find out what length side LN should be.

(b) Now draw the triangle accurately with a ruler and set square.
Measure the length of LN and see if it agrees with the length you calculated.

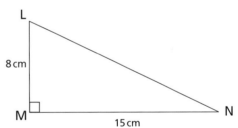

B7 Work out the missing lengths here.

(a)　(b)　(c)

(d)　(e)　(f)

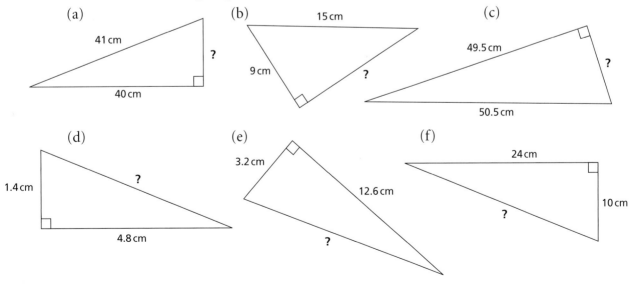

89

C Using square roots

In this right-angled triangle, $a = 8\,$cm and $b = 3\,$cm.

From the fact that $c^2 = a^2 + b^2$ it follows that

$$c^2 = 8^2 + 3^2$$
$$= 64 + 9$$
$$= 73$$

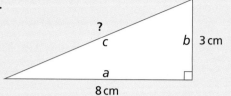

So c must be the **square root** of 73.

$$c = \sqrt{73} = \mathbf{8.54}\,\text{cm (to 2 d.p.)}$$

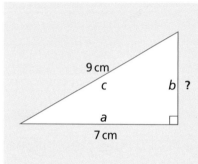

In this right-angled triangle, $a = 7\,$cm and $c = 9\,$cm but b is unknown.

From the fact that $a^2 + b^2 = c^2$

it follows that $7^2 + b^2 = 9^2$

so $49 + b^2 = 81$

so $b^2 = 81 - 49 = 32$

so $b = \sqrt{32} = \mathbf{5.66}\,\text{cm (to 2 d.p.)}$

C1 Work out the missing lengths. Give your answers to one decimal place.

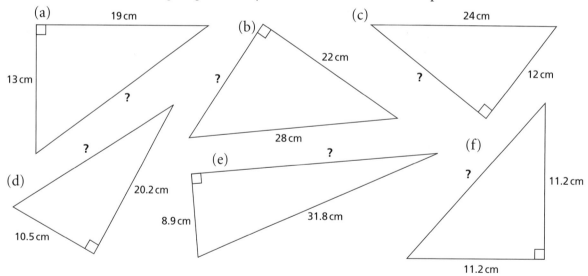

C2 (a) Use Pythagoras to work out side PQ to the nearest 0.1 cm.

(b) Now draw the triangle accurately with a ruler and set square.
Measure the length of PQ and see if it agrees with the length you calculated.

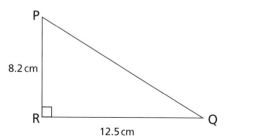

D Using Pythagoras

D1 A certain exercise book is 14 cm wide by 20 cm high.

(a) How long is the longest straight line you can draw on a single page?

(b) How long is the longest straight line you can draw on a double page?

D2 Measure the height and width of your own exercise book.
Repeat the calculations in D1 for your own book.
Measure to check your answers.

D3 This is the plan of a rectangular field.
There is a footpath across the field from A to C.
How much shorter is it to use the footpath
than to walk from A to B and then to C?

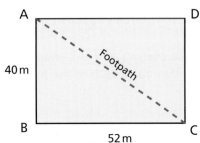

D4 Points A and B are plotted on a grid
on centimetre squared paper.

(a) Calculate the distance between A and B.

(b) Check by measuring the diagram.

(c) How long would a straight line
from (2, 2) to (14, 7) be?
(Draw this on a grid or make
a sketch if you need to.)

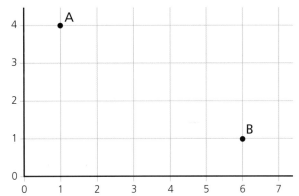

D5 (a) Calculate the lengths of the sides of
this quadrilateral.

(b) Use your working to say whether it is
exactly a rhombus.

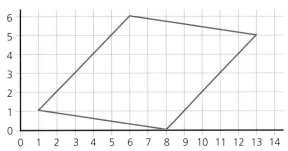

D6 Use Pythagoras to decide whether
this quadrilateral is exactly a kite.
Show your working.

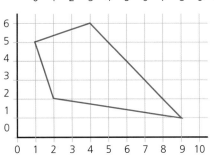

D7 Find by calculation which two of these points are closest together and which two are furthest apart.

        A (42, 11)   B (28, 36)   C (57, 34)

D8 How long is a straight line joining each pair of points if they are plotted on a centimetre square grid? Give your answers to two decimal places.

(a) (1, 3) to (5, 7)   (b) ($^-$2, 6) to (2, 8)   (c) (3, 1) to ($^-$6, 4)

(d) (6, 3) to (2, $^-$4)   (e) (11, $^-$1) to (8, 3)   (f) ($^-$7, 5) to ($^-$4, $^-$5)

(g) ($^-$2, $^-$4) to (5, $^-$6)   (h) (3, 3) to ($^-$2, $^-$1)   (i) ($^-$4, 2) to (5, $^-$3)

D9 All the corners of this L-shaped field are right angles. Calculate the length of the footpath.

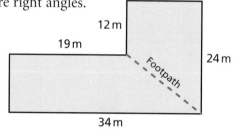

D10 A helicopter flies 26 km north from a heliport, then 19 km west. How far is it from the heliport now?

D11 A circular glass table has four thin legs. The legs are at the edge of the table top and at the corners of a square shelf under the top.

The diameter of the circular top is 1 metre.

(a) How long are the sides of the square shelf?

(b) What is the area of the square shelf?

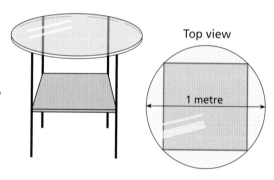

D12 Ali typed these instructions in a computer maze game.

How far is it, as the crow flies, from his finishing point back to his starting point?

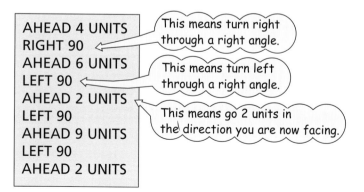

D13 Calculate the length of the fourth side of this quadrilateral.

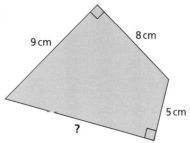

*D14 Say whether each of the angles in the triangle sketched here is equal to a right angle, less than a right angle or greater than a right angle.
Show your working.

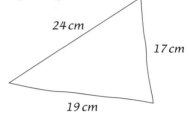

E Proving Pythagoras

- Trace this right-angled triangle four times, label the side lengths a, b, c and cut the triangles out.

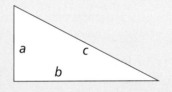

- Arrange the four triangles on this square so a pink square with area c^2 is showing.

- Now arrange the four triangles so a pink square with area a^2 and a pink square with area b^2 are showing.

- What does this show and why?
- Would this still work if lengths a and b changed?
  (Assume you could change the size of the pink square so its sides were still $a + b$ long.)

What progress have you made?

Statement

Evidence

I can work out the length of
a side of a right-angled triangle
if I know the lengths of the other two.

1 Find the missing lengths (to 1 d.p.).

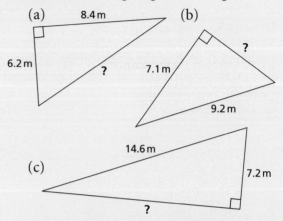

(a)   8.4 m   (b)

6.2 m   ?   7.1 m   ?

9.2 m

14.6 m

(c)   7.2 m

?

I can solve problems involving
the lengths of sides of right-angled
triangles.

2 This is a design for an ear-ring.
   All the angles in the outline are right angles.
   Calculate the lengths of the dotted
   lines (to 2 d.p.).

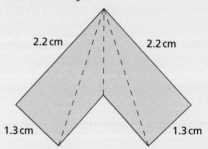

2.2 cm   2.2 cm

1.3 cm   1.3 cm

3 This trapezium has two right angles.

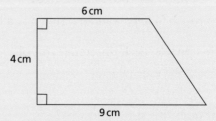

6 cm

4 cm

9 cm

Calculate (to 1 d.p.)
(a) the length of the fourth side
(b) the length of each of its diagonals

14 Simultaneous equations

This work will help you solve simultaneous equations using algebra.

A What's the difference?

A1 All weights in these balance
pictures are in grams.

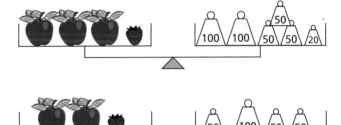

(a) Find the weight of an apple.

(b) What is the weight of a strawberry?

A2 All weights in these balance
pictures are in grams.

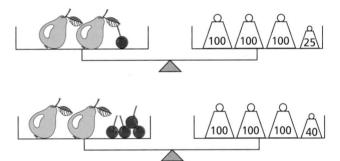

(a) Find the weight of a cherry.

(b) What is the weight of a pear?

A3 In a café, two teas and a bun cost £1.90.
Two teas and five buns cost £3.10.

How much is a tea and how much is a bun?

A4 In this puzzle, each different symbol
stands for a number.

$$■ + ♥ + ■ + ♥ + ♥ = 34$$

$$♥ + ■ + ◆ + ◆ = 14$$

$$♥ + ■ + ■ = 14$$

(a) What does each symbol stand for?

(b) Make up some puzzles like this for
someone else to solve.

B Addition and subtraction crosses

These are addition crosses.

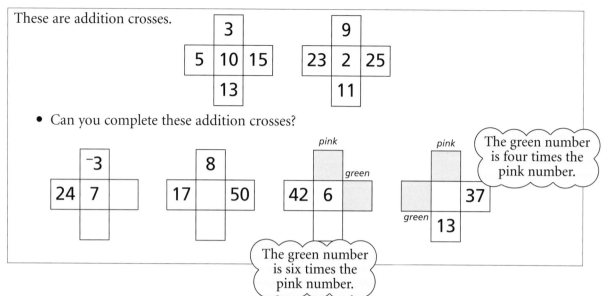

- Can you complete these addition crosses?

B1 This is an addition cross.

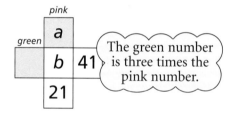

The green number is three times the pink number.

(a) Find an expression for the number in the green square.

(b) Write down two equations that involve *a* and *b*.

(c) Find the values of *a* and *b*.
Show all the numbers in the complete addition cross.

B2 Use algebra to find the missing numbers in these addition crosses.

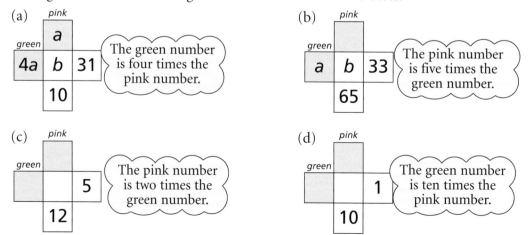

B3 (a) Copy and complete the last line of this working.

$$5a + 2b = 20$$
$$2a + 2b = 14$$

So $3a = $ .....

(b) Find the value of a.

(c) Use the first equation to find the value of b.

(d) Use the second equation to check that your values for a and b are correct.

B4 For each pair of equations, use algebra to find the values of a and b.

(a) $3a + b = 13$
$5a + b = 21$

(b) $6a + 3b = 21$
$4a + 3b = 23$

(c) $3a + 2b = 10$
$2b + 4a = 12$

(d) $a + 3b = 12$
$7b + a = 24$

(e) $2a + b = 9$
$2a + 6b = 4$

(f) $3a + 8b = 5$
$2b + 3a = 17$

Subtraction

B5 (a) Copy and complete the last line of this working.

$$a - b = 2$$
$$5a - b = 14$$

So $4a = $ .....

(b) Find the value of a.

(c) Use the first equation to find the value of b.

(d) Use the second equation to check that your values for a and b are correct.

B6 These are **subtraction** crosses.

Copy and complete the unfinished crosses.

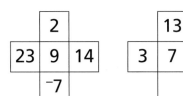

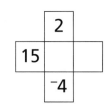

B7 This is a subtraction cross.

(a) Find an expression for the number in the green square.

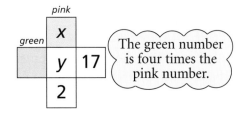

The green number is four times the pink number.

(b) Write down two equations that involve x and y.

(c) Find the values of x and y.
Show all the numbers in the complete subtraction cross.

B8 Use algebra to find the missing numbers in these subtraction crosses.

(a)

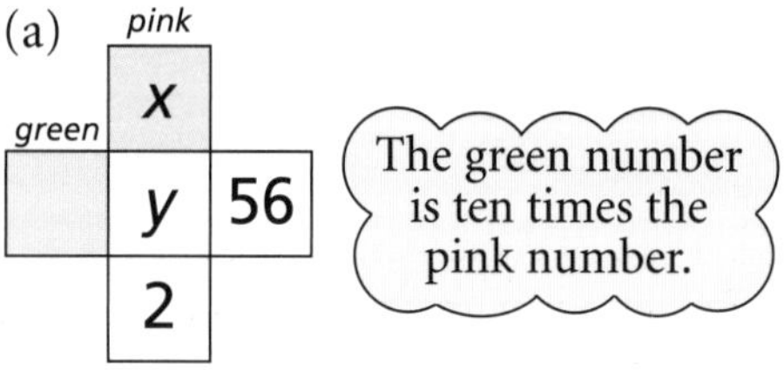
pink
green
x
y
56
2
The green number is ten times the pink number.

(b)

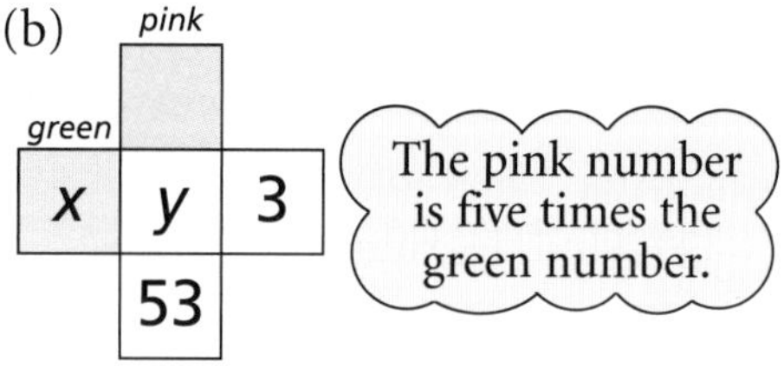
pink
green
x
y
3
53
The pink number is five times the green number.

(c)

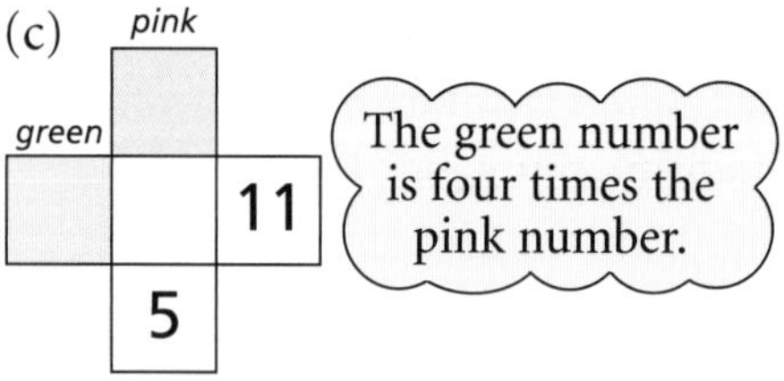
pink
green
11
5
The green number is four times the pink number.

(d)

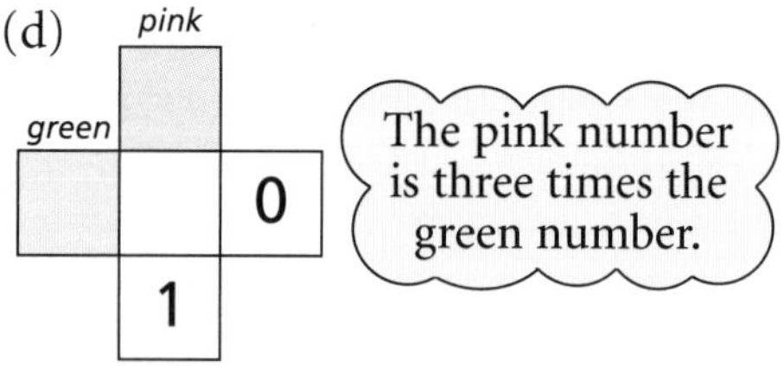
pink
green
0
1
The pink number is three times the green number.

B9 (a) Copy and complete the last line of this working.

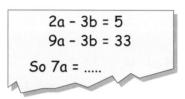

$2a - 3b = 5$
$9a - 3b = 33$

So $7a =$ .....

(b) Find the value of *a*.

(c) Use the first equation to find the value of *b*.

(d) Use the second equation to check that your values for *a* and *b* are correct.

B10 For each pair of equations, use algebra to find the values of *p* and *q*.

(a) $4p - q = 14$
$3p - q = 10$

(b) $5p - 2q = 29$
$p - 2q = 1$

(c) $2p - 4q = 6$
$3p - 4q = 11$

(d) $7p - 2q = 2$
$10p - 2q = 8$

(e) $p - q = 5$
$8p - q = 61$

(f) $6p - q = 32$
$3p - q = 17$

***B11** (a) What is the missing number in this **multiplication** cross?

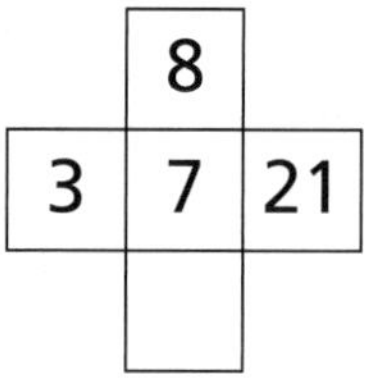
8
3
7
21

(b) Use algebra to find the missing numbers in these multiplication crosses.

(i)

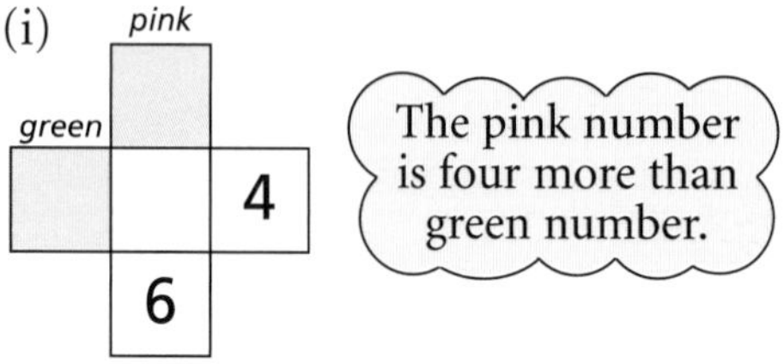
pink
green
4
6
The pink number is four more than green number.

(ii)

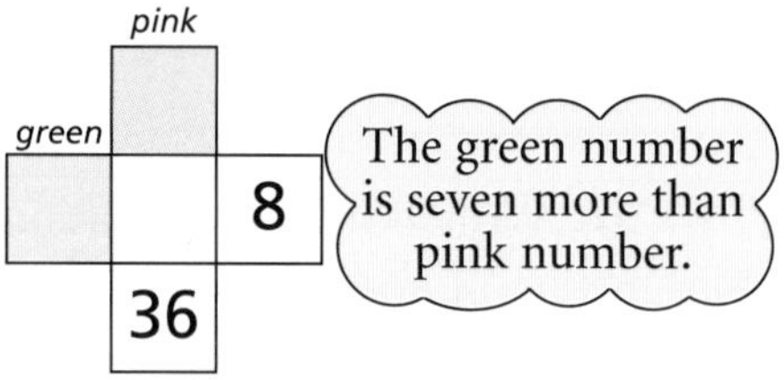
pink
green
8
36
The green number is seven more than pink number.

C Spot the difference

Patsy and Quentin each give Anna a number.

She multiplies Patsy's number by 3 and adds Quentin's number.

Her result is 40.

She also multiplies Patsy's number by 3 and subtracts Quentin's number.

This result is 32.

- What were Patsy and Quentin's numbers?

C1 (a) What do you need to add to ^-3y to give $4y$?

(b) Use your answer to help you solve this pair of simultaneous equations.

$$5x - 3y = 4$$
$$5x + 4y = 18$$

C2 (a) What do you need to add to ^-3y to give ^-y?

(b) Use your answer to help you solve this pair of simultaneous equations.

$$2x - y = 17$$
$$2x - 3y = 11$$

C3 Solve each pair of simultaneous equations.

(a) $2p - q = 8$
    $2p + 3q = 16$

(b) $3a + 2b = 18$
    $3a - 3b = 3$

(c) $4c - 2d = 6$
    $4d + 4c = 24$

(d) $3x - 4y = 14$
    $3x - y = 17$

(e) $5h - 2k = 19$
    $5h - 7k = 4$

(f) $m + n = 7$
    $m - n = 13$

C4 (a) David's and Nicola's ages add to give 17.
    If David is d years old and Nicola is n years old,
    write down an equation about David's and Nicola's ages.

(b) Nicola is older than David and the difference between their ages is 5 years.
    Write down another equation about David's and Nicola's ages.

(c) Solve these equations to find the ages of David and Nicola.

C5 Ahmet and Baljeet each think of a number (call them a and b).

Subtracting twice Baljeet's number from Ahmet's gives 9.
Adding three times Baljeet's number to Ahmet's gives 24.

Form two equations and solve them to find the values of a and b.

C6 If you add Cindy's age to twice Daniel's age you get 32.
If you take Cindy's age away from twice Daniel's you get 16.

How old is Cindy?

D Shall I compare thee...?

£2.75

£1.45

£4.40

£3.30

D1 Here is a pair of equations.

$$a + 5b = 47$$
$$3a + 8b = 85$$

(a) Multiply both sides of the first equation by 3 to give another equation.

(b) Find the values of a and b that fit both equations.

D2 Here is a pair of equations.

$$3m + 2n = 15$$
$$9n + 2m = 56$$

(a) Multiply both sides of the first equation by 2.

(b) Multiply both sides of the second equation by 3.

(c) Solve this pair of simultaneous equations.

D3 Here is a pair of equations.

$$2x + 5y = 14$$
$$5x + 4y = 18$$

(a) (i) Multiply both sides of the first equation by 4.

(ii) Multiply both sides of the second equation by 5.

(iii) Solve this pair of simultaneous equations.

(b) Now solve this pair of simultaneous equations by multiplying the first equation by 5 and the second by 2.

Make sure you get the same results as before.

D4 Solve each pair of simultaneous equations.

(a) $2v + w = 12$
$3v + 4w = 23$

(b) $5p + 4q = 41$
$p + 3q = 28$

(c) $3m + 2n = 21$
$2m + 11n = 72$

(d) $7k + 4h = 31$
$4k + 3h = 17$

(e) $3y + 4x = 28$
$7x + 9y = 64$

(f) $6a + 2b = 14$
$3b + 3a = 3$

Compare the way you solved each pair with others' ways.

D5 Julie and Rosa buy some candy bars and ice creams.

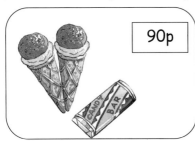

Julie spends 90p on two ice creams and a candy bar.
Rosa spends £2.05 on three ice creams and five candy bars.

With the cost of a candy bar *c* pence and the cost of an ice cream *i* pence,
form two equations and solve them to find the cost of a candy bar.

D6 Three geese and five golden eggs weigh 14 kg.
Five geese and three golden eggs weigh 18 kg.

Find the weight of a golden egg and the weight of a goose.

D7 Here is a pair of equations.

$$2x - 5y = 6$$
$$5x + 3y = 46$$

(a) Multiply both sides of the first equation by 5.

(b) Multiply both sides of the second equation by 2.

(c) Solve this pair of simultaneous equations.

D8 Here is a pair of equations.

$$5a - 3b = 13$$
$$7a - 4b = 19$$

(a) Multiply both sides of the first equation by 4.

(b) Multiply both sides of the second equation by 3.

(c) Solve this pair of simultaneous equations.

D9 Solve each pair of simultaneous equations.

(a) $2p - q = 8$
$5p - 3q = 19$

(b) $a - 2b = 1$
$4a - b = 25$

(c) $3d + c = 3$
$5c - 4d = 34$

(d) $4x - 2y = 6$
$5x + 3y = 35$

(e) $7h - 3k = 23$
$9h - 5k = 29$

(f) $2n + 7m = 8$
$2m - 5n = 19$

D10 Simonetta and Tomaso both choose a number.
If you add twice Simonetta's number to Tomaso's, you get 10.
If you take Simonetta's number away from three times Tomaso's number, you get 9.

Form two equations and solve them to find the numbers they chose.

E Useful arrangements

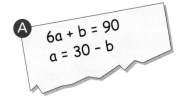

A
$6a + b = 90$
$a = 30 - b$

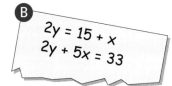

B
$2y = 15 + x$
$2y + 5x = 33$

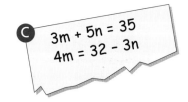

C
$3m + 5n = 35$
$4m = 32 - 3n$

E1 Solve each pair of simultaneous equations.

(a) $y = 6 - 2x$
$2x + 3y = 14$

(b) $5m = 27 + n$
$m - n = 7$

(c) $a = 3b + 1$
$a - b = 13$

(d) $2p = 4q + 2$
$2p + q = 17$

(e) $3y + 5z = 9$
$y = 11 - 5z$

(f) $5t = 41 + 3s$
$2t - 3s = 11$

E2 Solve each pair of simultaneous equations.

(a) $x = 47 - 3y$
$5x + 2y = 40$

(b) $2p + 3q = 9$
$3p = 14 - 4q$

(c) $2m = 17 - 3n$
$5n = 43 - 7m$

(d) $4x = y + 15$
$3x - 2y = 5$

(e) $5g = 2h + 31$
$3h = 1 - 2g$

(f) $2a = 3b + 10$
$9a = 5b + 96$

E3 An ornamental gourd seller sells golden and green gourds to tourists.
3 golden gourds and 9 green gourds cost 81 cents.
A golden gourd costs three times as much as a green one.

Find the price of each gourd.

What progress have you made?

Statement

I can solve simultaneous equations.

Evidence

1 Solve each pair of simultaneous equations.

(a) $2a + 4b = 18$
$2a + 2b = 14$

(b) $9x - y = 8$
$12x - y = 14$

(c) $4m - 2n = 4$
$n + 4m = 16$

(d) $2p + q = 11$
$5p + 3q = 27$

(e) $3v + 2w = 3$
$3w + 5v = 6$

(f) $6h - 4k = 7$
$4h - 6k = 3$

I can solve problems by forming and
solving simultaneous equations.

2 In mythical Zambania, there are two
different coins: dolas and kwatros.

8 dolas and 5 kwatros weigh 235 grams.
4 dolas and 6 kwatros weigh 170 grams.

Find the weight of each coin.

Review 2

1 The graph of $y = x^3 + x$ crosses the line $y = 4$ between $x = 1.2$ and $x = 1.4$.

It follows that the equation $x^3 + x = 4$ has a solution between 1.2 and 1.4.

Use trial and improvement to find this solution correct to two decimal places.

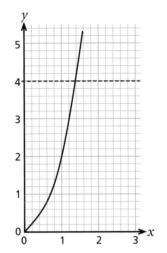

2 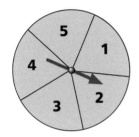 This spinner is spun twice.
Work out the probability that

  (a) both spins give odd numbers

  (b) both spins give even numbers

  (c) one of the spins gives an odd number and the other even

3 Draw axes with x from 0 to 6 and y from $^-4$ to 4.

  (a) Draw the graph of the equation $x + 2y = 4$.

  (b) By drawing the graph of $2x - 2y = 5$, find the values of x and y that satisfy the simultaneous equations $x + 2y = 4$ and $2x - 2y = 5$.

4 Why does this chart give a misleading picture of the growth in the membership of a sports club?

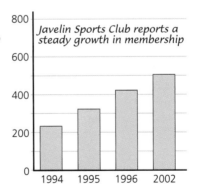

5 Calculate, to the nearest 0.1 cm, the lengths marked with letters.

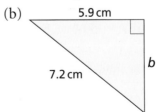

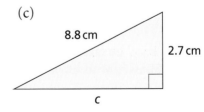

103

6 Solve each of these pairs of simultaneous equations.

(a) $3x + y = 9$
$x - y = 1$

(b) $2x + 3y = 13$
$x + 2y = 9$

7 Karen plays two fairground games, 'Roll a penny' and 'Darts'.

The probability that she wins at 'Roll a penny' is $\frac{1}{5}$.
The probability that she wins at 'Darts' is $\frac{1}{6}$.
The outcomes of the two games are independent.

(a) Copy and complete the tree diagram.

(b) Find the probability that Karen

    (i) wins at both games

    (ii) wins at one game and loses at the other

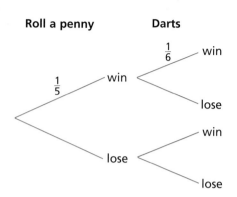

8 The time taken by a computer printer to print a document is proportional to the number of sheets in the document.

If the printer takes 24 minutes to print 60 sheets, how long will it take to print 80 sheets?

9 Calculate the value of x.

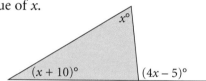

10 This diagram shows the end wall of a building.

Calculate

(a) the length marked p

(b) the area of the wall

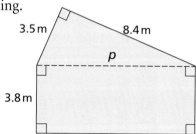

11 Without using a calculator, change each of these fractions to a decimal.

(a) $\frac{13}{50}$ (b) $\frac{17}{20}$ (c) $\frac{3}{8}$ (d) $\frac{5}{16}$

12 Brad was asked to work out the mean of these numbers

    575   579   573   577   574   578

'Easy,' he said, 'I can do it in my head.'
Show how to work out the mean of these numbers in your head.

13 The circumference of a circle is 45.2 cm.

Calculate to 1 d.p.   (a) the radius in cm   (b) the area in cm^2

14 The purple graph shows an express train passing through Breading Station.
The red graph shows another train that stops at Breading.

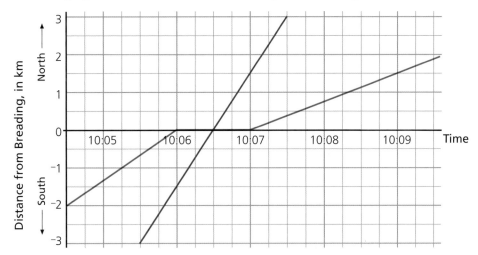

(a) For how long does the stopping train stop at Breading?

(b) What is the speed of the express train, in km per minute?

(c) Convert the speed of the express train to km per hour.

(d) What is the speed, in km per hour, of the stopping train

    (i)   as it approaches Breading Station

    (ii) as it leaves Breading Station

(e) The next station going north from Breading is 9 km from Breading.
If the stopping train continues to travel at the speed with which it
left Breading, at what time will it arrive at the next station?

15 Solve each of these equations.

(a) $18 - 3x = 5x + 8$ (b) $3(2x - 5) = x - 20$ (c) $40 - (7 + 2x) = x$

(d) $5x - 2(x - 4) = 50$ (e) $\dfrac{5x + 6}{3} = x$ (f) $\dfrac{2x - 9}{5} = x + 3$

16 A supermarket was selling canned peaches in 500 g cans for 80p each.
It then changed the size of the can to 400 g, but still charged 80p per can.

(a) Calculate the price per kilogram before and after the change.

(b) Calculate the percentage increase in the price per kilogram.

*17 Graham had 8 litres of orange juice and Jane had 5 litres.
Graham gave some of his juice to Jane. Afterwards he had half as much as Jane.

How much did Graham give Jane?

⑮ Spot the errors

This work will help you check working and make sure that answers are sensible.

In every one of these pieces of work there is at least one error.
The working out may be wrong, or the answer may not be sensible.
Spot the errors and correct them.

1 Calculate $\frac{78}{23 \times 1.9}$ correct to two decimal places.

> 78 ÷ 23 x 1.9 = 6.44 to 2 d.p.

2 138 children are to have school lunch.

Each table seats 8 children.

How many tables are needed?

> Number of tables = 138 ÷ 8 = 17.25
> = 17 (to the nearest whole number)

3 Calculate the length marked c in this right-angled triangle.

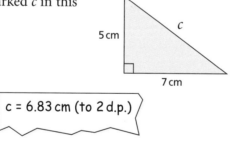

5 cm

c

7 cm

> c = 6.83 cm (to 2 d.p.)

4 A ferry travels 125 miles at an average speed of 21 m.p.h.
How long does the journey take?

> Time taken = 125 ÷ 21 = 5.9523809
> = 5 hours 95 minutes

5 A car travels 210 miles in 4 hours 30 minutes.
  Calculate its average speed.

> Average speed = 210 ÷ 4.30 = 48.837209 m.p.h.

6 The exchange rate between £ and US$ is £1 = $1.63.
  Change $5.70 to £.

> 5.70 × 1.63 = 9.291
> Answer £9.291

7 The volume of this cuboid is 107.5 cm³.
  Calculate its length.

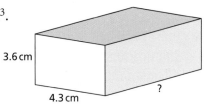

3.6 cm

4.3 cm

?

> Area of end = 3.6 × 4.3 = 15.48 = 15 cm²
> Length = 107.5 ÷ 15 = 7.16 = 7 cm

8 A sheet of paper is 0.2 m by 0.3 m.
  Calculate its area in cm².

> Area = 0.2 × 0.3 = 0.6 m²
> Change this to cm²: 0.6 × 100 = 60 cm²

*9 A cyclist travels the 42 miles from Ayton to Beaham in 4 hours.
  He does the return journey in $5\frac{1}{2}$ hours.
  Calculate his average speed for the whole trip, there and back.

> Average speed A to B = 42 ÷ 4 = 10.5 m.p.h.
> Average speed B to A = 42 ÷ 5.5 = 7.64 m.p.h. (to 2 d.p.)
> Average speed for whole trip = $\frac{10.5 + 7.64}{2}$ = 9.07 m.p.h.

⓰ Changing the subject

This work will help you

♦ substitute numbers into a formula and solve the resulting equation

♦ rearrange a formula to make a different letter the subject

A From formula to equation

A1 Towers like this are made from 1 metre bars.

The number N of bars in a tower of height h metres is given by the formula

$$N = 8h + 4$$

(a) Find the value of N when $h = 7$.

(b) If you are told that $N = 28$, then the value of h is given by the equation $28 = 8h + 4$.

Solve this equation to find the value of h.

(c) Find the value of h for which

    (i) $N = 76$ (ii) $N = 140$

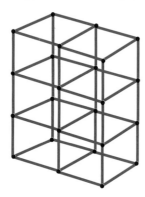

A2 The formula for the number of bars in frameworks like the one on the left is

$$N = 13h + 7$$

Find the value of h when

(a) $N = 46$

(b) $N = 163$

(c) $N = 189$

A3 (a) Find the formula for the number of bars in frameworks like this.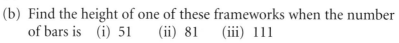

(b) Find the height of one of these frameworks when the number of bars is   (i) 51   (ii) 81   (iii) 111

(c) If you have 130 bars, what is the highest framework of this type that you can make?

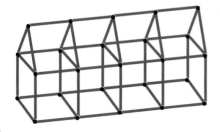

A4 (a) Find the formula for the number N of bars in a framework like the one on the left, of length l metres.

(b) Find the value of l when N is   (i) 105   (ii) 160

B Reversing the flow

The formula $N = 8h + 4$ can be shown as a flow diagram.

Reversing the diagram gives a formula for h in terms of N.

$$h = \frac{N-4}{8}$$

B1 (a) Draw the flow diagram for the formula $N = 13h + 7$.

(b) Find a formula for h in terms of N.

B2 (a) Copy and complete this flow diagram for the formula $s = 2(p - 5)$.

(b) By reversing the diagram, find a formula for p in terms of s.

B3 (a) This is the flow diagram for a formula.
Write down the formula.

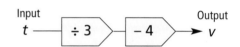

(b) Reverse the flow diagram and write down
the formula for the reversed diagram.

B4 The variables m and r are connected by the formula $m = \dfrac{r-7}{6}$.
Find a formula for r in terms of m.

B5 The variables b and a are connected by the formula $b = \dfrac{a}{2} + 9$.
Find a formula for a in terms of b.

B6 In each case below, find the formula for x in terms of y.

(a) $y = 3x - 2$ (b) $y = 3(x - 2)$ (c) $y = \dfrac{x+4}{5}$ (d) $y = \dfrac{x}{3} - 8$

B7 (a) Copy and complete this flow diagram
for the formula $y = \dfrac{3(x-2)}{5}$.

(b) By reversing the diagram, find a formula for x in terms of y.

B8 (a) Draw the flow diagram for the formula $y = \dfrac{2x+5}{4}$.

(b) Find a formula for x in terms of y.

B9 Given that $y = x^2 + 5$, find a formula for x in terms of y.

C Squares and square roots

The inverse of the operation square is square root.

Here is the flow diagram for $y = 3x^2 - 4$.

The reversed diagram gives the formula

$$x = \sqrt{\frac{y+4}{3}}$$

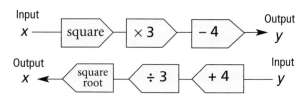

Strictly speaking, we should write $\pm\sqrt{}$, because every number greater than 0 has **two** square roots, one positive and the other negative.

For example, there are two numbers whose square is 9; they are 3 and ⁻3.

However, in many cases only a positive result makes sense.

C1 The surface area S of a cube of edge length l is given by the formula $S = 6l^2$.

    (a) Draw a flow diagram for this formula.

    (b) By reversing the diagram, find the formula for l in terms of S.

C2 Find the formula for u in terms of v in each case below.

    (a) $v = u^2 + 4$ (b) $v = (u - 2)^2$ (c) $v = \dfrac{u^2 + 7}{3}$ (d) $v = 5u^2 + 3$

    (e) $v = \sqrt{u} - 3$ (f) $v = \sqrt{2u - 1}$ (g) $v = (u + 5)^2 + 2$ (h) $v = 2(u - 3)^2$

D Formulas with several letters

In the formula $u = 3s - 5$, u is called the **subject** of the formula.

When the formula is rewritten as $s = \dfrac{u+5}{3}$, the subject has been changed to s.

If a formula has several letters, there will be different ways to rearrange the formula and change the subject.

Worked example

Make p the subject of the formula $q = ap - b$.

Think of p as the input in the formula.
Show what is done to p in a flow diagram.

Reverse the flow diagram.
Write down the corresponding formula.

$$p = \frac{q+b}{a}$$

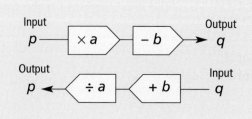

D1 The perimeter P of a rectangle of length l and width w is given by the formula
$$P = 2(l + w)$$

(a) Thinking of l as the input, draw a flow diagram for finding P.

(b) By reversing the flow diagram, find the formula for l in terms of P and w.

(c) Use your result to calculate l when $P = 18$ and $w = 5$.

D2 Rearrange each of these formulas so that the letter shown in red is the subject.

(a) $y = ax + b$ (b) $q = k(p + 4)$ (c) $s = a(u - b)$ (d) $y = \dfrac{x}{d} - b$

(e) $w = \dfrac{u + a}{b}$ (f) $z = a\left(\dfrac{t}{r} + p\right)$ (g) $y = \dfrac{ax - b}{c}$ (h) $q = \dfrac{k(p + h)}{m}$

D3 The cross-section of this prism is a square of side a.
The length of the prism is l.

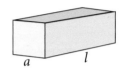

(a) The volume V of the prism is given by the formula $V = a^2 l$.
Make a the subject of this formula.

(b) The surface area S of the prism is given by the formula $S = 4al + 2a^2$.
Make l the subject of this formula.

D4 The power P watts consumed in an electrical component is related to the voltage V volts and the the resistance R ohms by the formula

$$P = \frac{V^2}{R}$$

Make V the subject of this formula.

*D5 The volume V of a sphere of radius r is given by the formula $V = \dfrac{4\pi r^3}{3}$
Make r the subject of this formula.

What progress have you made?

Statement		Evidence
I can change the subject of a formula containing two letters.	1	Make x the subject of each of these formulas. (a) $y = \dfrac{x}{4} + 5$ (b) $y = \dfrac{5x - 2}{3}$
I can change the subject of a formula containing more than two letters.	2	Make p the subject of each of these formulas. (a) $q = \dfrac{p - r}{s}$ (b) $q = a\left(\dfrac{p}{b} + c\right)$
I can change the subject of a formula involving a square.	3	The surface area S of a sphere of radius r is given by the formula $S = 4\pi r^2$. Make r the subject of this formula.

17 Similar shapes

This is about shapes that are scaled copies of one another.
The work will help you

◆ use scale factors
◆ use ratios in problems about similar shapes

A Scaling

A graphic artist has designed a logo
for a company.

He is asked to make a copy double
the size.
But he isn't happy with his drawing.

What is wrong with it?

Original

Copy

Here is part of another copy of the original logo.
This one is correctly scaled.
How high is the complete copy?

?

A1 The larger picture here is meant to be a scaled copy of the original.

 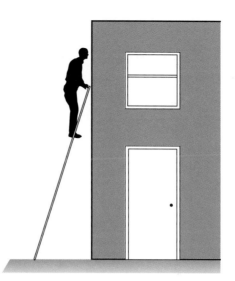

(a) Copy and complete this table of measurements and work out each multiplier.

Measurement	Original length	× ?	Length in copy
Height of building			
Length of ladder			
Height of door			
Width of door			

(b) What is the scale factor of the enlargement?

(c) Measure the angle between the ladder and the ground in the original and the copy. What do you find?

A2 (a) Measure the base and height of this original isosceles triangle.

(b) Measure the base and height of each of the other isosceles triangles.

(c) For each of the other triangles find out

  (i) what the original base has been multiplied by to give its base

  (ii) what the original height has been multiplied by to give its height

(d) Which of the triangles are scaled copies of the original?

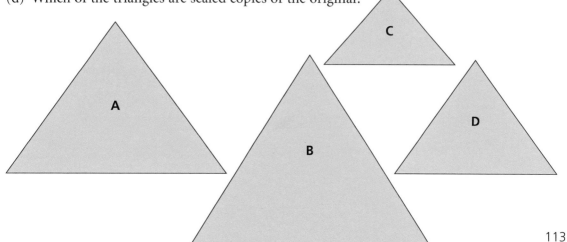

113

A3 A picture shows a house, a car and a hedge.
This table compares measurements in the original picture and a scaled copy.
Copy the table and fill in the missing values.

	Original length	Scale factor	Copy length
Width of picture	10 cm		25 cm
Height of picture	6 cm		
Height of house	3 cm		
Length of car			5 cm
Length of hedge			20 cm

A4 Bella has a rectangular photo 125 mm by 80 mm.
The photo is enlarged. The longer side of the copy is 350 mm long.

Calculate the length of the shorter side of the copy.

A5 Karl has a picture of himself standing by a tree outside his house.
He makes three enlarged copies A, B, C of the picture.

(a) Use the measurements given in this table to work out the scale factor
of each enlargement.

Measurement	Original length	Length in copy A	Length in copy B	Length in copy C
Karl's height	20 mm	25 mm		
Height of house	120 mm		210 mm	
Height of tree	150 mm			360 mm

(b) Calculate the missing measurements in the table.

B Scaling down

This pencil is life size.

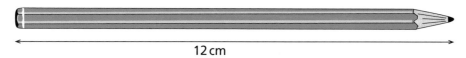

12 cm

In the copy below it is $\frac{1}{4}$ of its original size. The scale factor is $\frac{1}{4}$.

3 cm

B1 What is the scale factor of each of these copies?

(a)

(b)

(c)

(d)

Sometimes it is easier to use decimals.

10cm

Original

3.5cm

copy

The scale factor is copy length ÷ original length = 3.5 ÷ 10 = **0.35**

B2 Find the scale factor for each of these copies of the original lizard.

(a)

(b)

(d)

(c)

B3 Copy and complete this table for a picture that has been copied.

	Original length	Scale factor	Copy length
Width of picture	8 cm		4.8 cm
Height of picture	5 cm		
Height of tree	3 cm		
Length of pond			4.5 cm
Length of fence			5.4 cm

B4 Angela has a picture of herself playing the flute.
She makes three scaled copies A, B, C of the picture.

Some of the measurements are shown in this table.

Measurement	Original length	Length in copy A	Length in copy B	Length in copy C
Angela's height	25 mm	20 mm		
Length of flute	10 mm		4 mm	
Height of music stand	20 mm			11 mm

(a) Find the scale factor for each copy.

(b) Work out the missing measurements.

C Ratios within shapes

Paper sizes

You need some A2 paper, broadsheet newspaper pages, metre rule.

You will often have used A4 paper. This is a standard metric size.
Newspapers, however, use a different system.

- Measure the lengths of the sides of the A2 sheet and of the broadsheet.
- Fold each sheet in half, by halving the longer side.
  Measure the sides of the new sheet.
- Repeat this another three times for each sheet.
- Record your results in a table and calculate the ratio $\dfrac{\text{long side}}{\text{short side}}$ each time.

	Metric sizes					Newspaper sizes				
	A2	A3	A4	A5	A6	Broadsheet	Tabloid	$\frac{1}{2}$T	$\frac{1}{4}$T	$\frac{1}{8}$T
long side										
short side										
$\frac{\text{long side}}{\text{short side}}$										

- What rule applies to the A sizes?
  Does the same rule apply to the newspaper sizes?
- What advantage does the A system have over the newspaper sizes?

A ratio is used to compare two quantities.

For example, if the height of a window is 3 times its width,
we can write

$$\text{height} = 3 \times \text{width}$$

$$\text{or} \quad \frac{\text{height}}{\text{width}} = 3$$

We say 'the ratio of height to width is 3 (or 3 : 1)'.

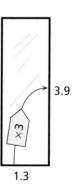

3.9

×3

1.3

For this window, the height is 0.4 times the width,

$$\text{so the ratio} \quad \frac{\text{height}}{\text{width}} = 0.4$$

1.2

3

C1 These two triangles are scaled copies of each other.

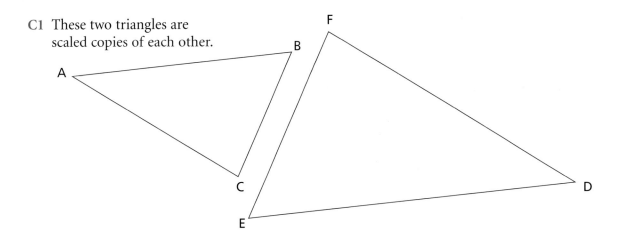

(a) Measure the lengths of AB and DE.
Use these lengths to find the scale factor of enlargement of DE from AB.

(b) Measure the lengths of BC and EF.
What is the scale factor of enlargement of BC from EF?

(c) Without measuring give the scale factor of CA to FD.

(d) Measure angle ABC. What can you say about angle DEF?

(e) Measure angle BCA. What can you say about angle EFD?

(f) Without measuring find angles CAB and FDE.

C2 For each of these find the ratio $\dfrac{\text{longer side}}{\text{shorter side}}$.

Use this to make pairs which are copies from the same original.

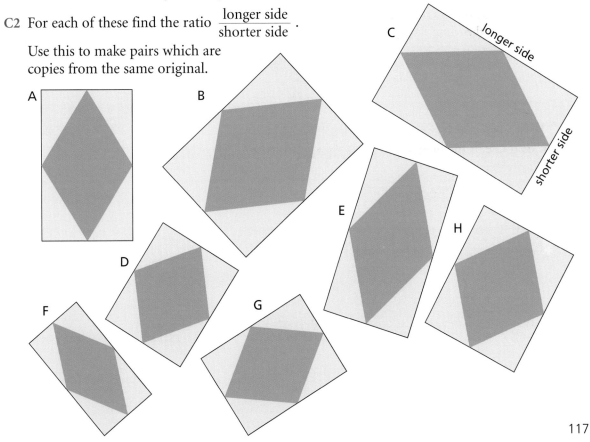

C3 This table shows measurements from scaled copies of the same original.
Copy and complete the table.

Height of picture	Ratio $\frac{width}{height}$	Width of picture
5 cm		8 cm
7.2 cm		
		12 cm

C4 Copy and complete this table for a set of scaled copies.

Height of picture	×? →	Width of picture
18 cm		12 cm
8.4 cm		
		20 cm

D Scale factors and ratios

Shield B is a scaled copy of shield A.

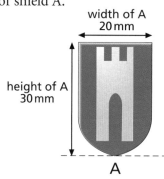

width of A
20 mm

height of A
30 mm

A

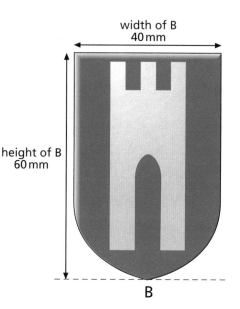

width of B
40 mm

height of B
60 mm

B

The scale factor compares shield B with shield A.

	shield B	÷	shield A	
height	60 mm	÷	30 mm	= 2
width	40 mm	÷	20 mm	= 2

Shield B is an enlargement of A with scale factor 2.

We can also compare the height and width of each individual shield.

	shield B		shield A	
$\frac{height}{width}$	$\frac{60 \text{ mm}}{40 \text{ mm}}$	=	$\frac{30 \text{ mm}}{20 \text{ mm}}$	= 1.5

The ratio $\frac{height\ of\ shield}{width\ of\ shield}$ will be 1.5 for **any** enlargement of shield A.

D1 Measure the height and the width of the tower on each shield.

Show that

(a) the ratios $\dfrac{\text{height of tower B}}{\text{height of tower A}}$ and $\dfrac{\text{width of tower B}}{\text{width of tower A}}$ are each 2

(b) the ratios $\dfrac{\text{height of tower A}}{\text{width of tower A}}$ and $\dfrac{\text{height of tower B}}{\text{width of tower B}}$ are equal

D2 (a) Find the ratio $\dfrac{\text{width of tower}}{\text{width of shield}}$ for each of the shields.

(b) Another shield, C, is also a scaled copy of shield A.
Shield C is 50 mm wide.

(i) How high is shield C? (ii) How wide is the tower on shield C?

(iii) How high is the tower on shield C?

E Similarity

When two shapes are exact scaled copies of one another, we say they are **similar**.
In mathematics, 'similar' has this precise meaning.

So if two shapes are similar,

* corresponding angles in the two shapes are equal
* the ratio of a pair of lengths in one shape is equal to the corresponding ratio in the other

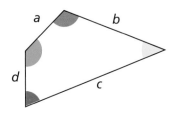

$$\frac{a}{b} = \frac{a'}{b'}$$

$$\frac{c}{a} = \frac{c'}{a'}$$

and so on

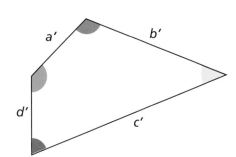

E1 By finding the ratio $\dfrac{\text{longer side}}{\text{shorter side}}$ say which of these rectangles are similar to P.

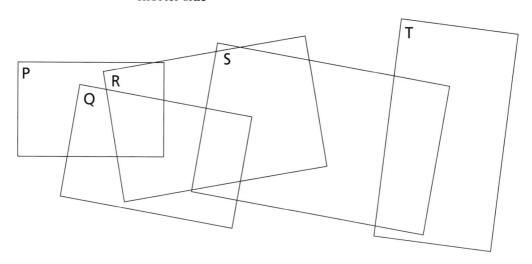

E2 These triangles are not drawn accurately.
Explain why, if they were drawn accurately, they would be similar.

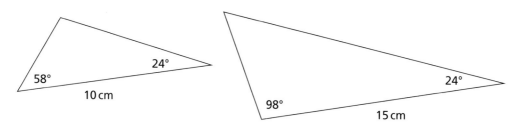

E3 Find the ratio $\dfrac{a}{b}$ in each of these right-angled triangles.
Use this to decide which are similar to the shaded one.

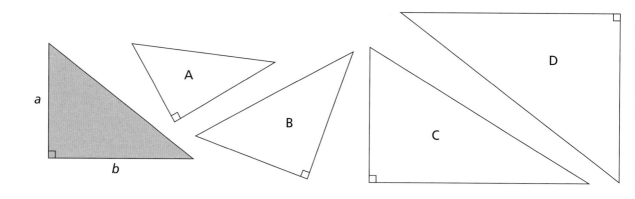

E4 (a) Calculate the ratio $\dfrac{\text{long side}}{\text{shorter side}}$ for the green rectangle below.

(b) Each of the rectangles A, B and C is similar to the green rectangle but only partly visible.

Calculate the longer side of each of them.

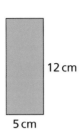

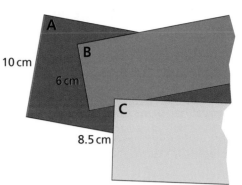

E5 These right-angled triangles are all similar to the grey triangle.

(a) What is the ratio $\dfrac{\text{height}}{\text{base}}$ for the grey triangle?

(b) Use this ratio to find the lengths labelled p, q, r, …

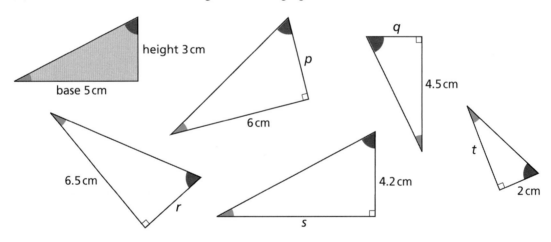

Finding the heights of tall objects

You need sheet 251.

Make up the ratiometer from the sheet.

Measure a suitable distance from a tall object.
Ask someone to check the ratiometer is level.

Sight the top of the object between the two Vs
by sliding the vertical bar up or down.

Find the ratio needed to obtain the height on the
vertical bar from the length (20 cm) to your eye.

The same ratio is used to find the height of your
object from the distance you are standing from it.

Use the ratiometer to find the height of some tall object.
For an accurate result you will need to add the height of your eyes above ground.

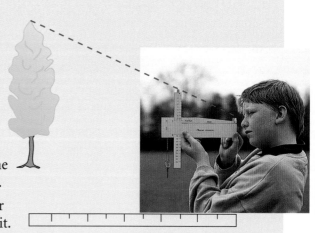

What progress have you made?

Statement

I can find the scale factor of a copy.

Original

I can find and use ratios in problems
about similar shapes.

Evidence

1 Find the scale factor used to make each of
these copies from the original.

A

B

2 These three triangles are similar
   (a) Find the ratio $\frac{p}{q}$ in the first triangle.

   (b) Calculate the length labelled r.

   (c) Calculate the length labelled s.

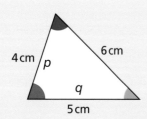

4 cm p 6 cm

q

5 cm

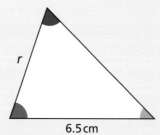

r

6.5 cm

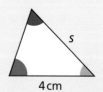

s

4 cm

⑱ Functions and graphs

This work will help you

◆ draw graphs of simple quadratic functions

◆ draw graphs of some other types of function

A Quadratic functions

- A **function** is a rule linking an input and an output.
  Here is an example of a function, written in three different ways.

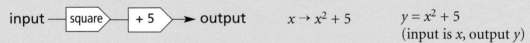

input — [square] — [+ 5] → output $x \to x^2 + 5$ $y = x^2 + 5$
          (input is x, output y)

- The function above is a **quadratic** function.
  Here are some other examples of quadratic functions, written in the form $y = \ldots$

$$y = 3x^2 \qquad y = 4x^2 - 3x \qquad y = x^2 + 3x - 1 \qquad y = 7 + 3x - 2x^2$$

A1 Draw axes on graph paper with x from ⁻3 to 3 and y from ⁻2 to 10.

(a) Copy and complete this table of values for the function $y = x^2$.

x	⁻3	⁻2.5	⁻2	⁻1.5	⁻1	⁻0.5	0	0.5	1	1.5	2	2.5	3
y	9	6.25											

(b) Plot the points from the table. Draw a smooth curve through them and label the graph '$y = x^2$'.

(c) Describe the symmetry of the graph.

(d) Use the graph to find, as accurately as you can, the values of x for which $y = 6$.

(e) Show how the graph can be used to find an approximation to $\sqrt{3}$.

A2 Here is a rough sketch of the graph of $y = x^2$.

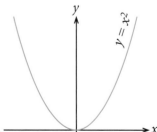

(a) What do you think the graph of $y = x^2 + 1$ will look like?
Copy the sketch of $y = x^2$ and add a sketch of $y = x^2 + 1$.

(b) Make a table of values for the function $y = x^2 + 1$ and draw the graph on the same set of axes that you used for question A1.

A3 (a) Copy and complete this table of values for the function $y = x^2 - 2$.

x	-3	-2	-1	0	1	2	3
y	7						

(b) Draw the graph of $y = x^2 - 2$ on the same axes as for question A1.

(c) Use the graph to find approximately the values of x for which $x^2 - 2 = 0$.

A4 The graph of $y = x^2$ is shown in black.

Write down the equation of the other graphs.

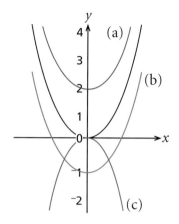

A5 You need sheet 252.

The graph of $y = x^2$ is already drawn on the sheet.

(a) Complete the table of values for $y = 2x^2$.
(Remember that $2x^2$ means 2 times (x^2).)

Draw the graph of $y = 2x^2$.

(b) Complete the table of values for $y = \frac{1}{2}x^2$.

Draw the graph of $y = \frac{1}{2}x^2$.

(c) Without making another table of values, draw the graph of

(i) $y = 2x^2 - 1$ (ii) $y = \frac{1}{2}x^2 + 3$

*A6** Find the equation of each of these graphs.

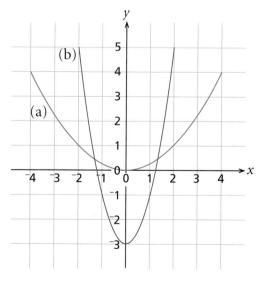

B Other functions

B1 You need sheet 253.

Complete the tables of values on the sheet and draw the graphs of these cubic functions.

$$y = x^3 \qquad y = x^3 + 5 \qquad y = x^3 + x$$

B2 (a) Copy and complete this table for the function $y = \dfrac{6}{x}$.

x	‾6	‾5	‾4	‾3	‾2	‾1	1	2	3	4	5	6
y	‾1	‾1.2	‾1.5									

(b) Plot the points on graph paper but do **not** join them up yet!

(c) What is y when x is

(i) 0.5   (ii) 0.1   (iii) 0.01

This should help you see what happens to the value of y for positive values of x getting closer and closer to 0.

(d) What is y when x is

(i) ‾0.5   (ii) ‾0.1   (iii) ‾0.01

This should help you see what happens to the value of y for negative values of x getting closer and closer to 0.

(e) Explain why there is no place in the table for $x = 0$.

(f) Draw the graph of $y = \dfrac{6}{x}$.

B3 This is a sketch of the graph of $y = \dfrac{-5}{x}$.

(a) What are the coordinates of point A?

(b) What are the coordinates of point B?

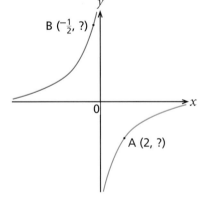

B $\left(-\tfrac{1}{2},\ ?\right)$

A (2, ?)

B4 Use a graph drawing program on a computer to investigate the shape of the graphs of

$$y = \frac{1}{x} \qquad y = \frac{1}{x^2} \qquad y = \frac{1}{x^3} \qquad \text{and so on}$$

What progress have you made?

Statement

Evidence

I can draw the graph of a simple quadratic function.

1 (a) Make a table of values for the function $y = 12 - x^2$, for values of x from $^-4$ to 4.

(b) Draw the graph of $y = 12 - x^2$.

(c) Use your graph to find approximately the values of x for which $12 - x^2 = 0$.

I can draw the graph of a simple cubic function.

2 (a) Make a table of values for the function $y = x^3 - 3$, for values of x from $^-3$ to 3.

(b) Draw the graph of $y = x^3 - 3$.

(c) On the same axes, show the graph of $y = x^3 + 3$.

I can draw the graph of a function of the form $y = \frac{A}{x}$.

3 (a) The graph of $y = \frac{12}{x}$ goes through these four points. Complete their coordinates.
$(3, \ldots)$ $(\ldots, {}^-3)$ $({}^-3, \ldots)$ $(4, \ldots)$

(b) Sketch the graph of $y = \frac{12}{x}$.

19 Inequalities

This work will help you

◆ understand and use simple inequalities

◆ combine simple inequalities

◆ convert statements in words to inequalities in symbols

A Simple inequalities

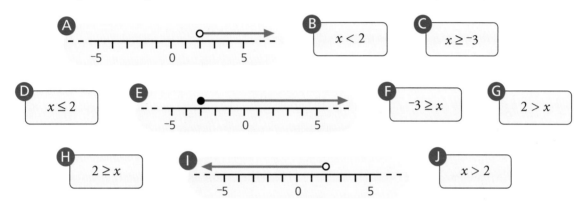

A1 Decide if each of the following is true or false.

(a) $1 > {}^-4$

(b) $^-2 > {}^-1$

(c) $\sqrt{2} < 2$

(d) $\pi \geq 4$

(e) $7 \leq 7$

(f) $\frac{1}{3} > 0.33$

(g) $0.19 \leq 0.2$

(h) $0.9^2 > 0.9$

(i) $\frac{1}{3} < \frac{1}{4}$

(j) $\pi + 2 > 5$

(k) $\sqrt{26} \geq 5$

(l) $\frac{5}{9} \geq \frac{6}{11}$

A2 (a) Which of the numbers in the bubble are in the set of values for x when $x \leq 1$?

(b) Which are in the set of values for n when $n > 3$?

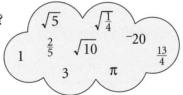

A3 Represent each inequality on a number line.

(a) $x < 4$

(b) $n \geq {}^-1$

(c) $5\frac{1}{2} > p$

(d) $^-3 \leq y$

A4 Write inequalities to describe the following diagrams.

(a)

(b)

(c)

(d)

A5 If we know that $x > 3$, is each of the following

- always true
- sometimes true
- never true

(a) $x > 4$ (b) $x > 2$ (c) $x > {}^-1$ (d) $x < 1$

(e) $x + 5 > 8$ (f) $x - 1 < 4$ (g) $2x > 3$ (h) $\frac{x}{2} < 1$

A6 If we know that $n \leq 1$, is each of the following

- always true
- sometimes true
- never true

(a) $3 < n$ (b) $n > 0.5$ (c) $1.5 \geq n$ (d) $n - 1 < 0$

(e) $5n + 1 < 10$ (f) $3n > 6$ (g) $\frac{n}{5} < 0.5$ (h) $\frac{5}{n} > 1$

A7 List five different values for m so that $m + 2 < 5$.

A8 Find four different positive values for n so that $5n \leq 6$.

A9 Find the largest prime number p so that $p < 100$.

Reminder An integer is a positive or negative whole number (including 0).
These are all integers: $7, {}^-5, 3, 0, {}^-1, {}^-15, 20$

A10 (a) Find all the integers n such that $n^2 < 10$. (Hint: there are seven of them.)

(b) Find all the integers n such that $n^2 < 20$.

(c) Find all the integers n such that $n^2 \leq 36$.

B Combined inequalities

The inequality $x \geq 1$ can also be written $1 \leq x$ and can be shown like this.

The inequality $x < 4$ can be shown like this.

If both inequalities are true, then x must be in the interval shown here.

We can write this interval as $1 \leq x < 4$.

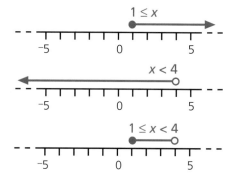

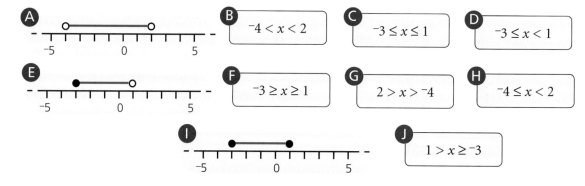

B1 Where possible, combine each of the following pairs of inequalities.

(a) $5 < x$ and $x < 10$ (b) $x > 6$ and $x < 1$ (c) $x < 3$ and $x \geq {}^-4$

(d) $x \geq 0$ and $9 > x$ (e) $x < 1$ and $x < 10$ (f) $12 \geq x$ and $^-9 \leq x$

(g) $^-5 \geq x$ and $x \geq 4$ (h) $x \geq 3$ and $^-3 < x$

B2 (a) Which of the numbers in the bubble are in the set of values for x when $0 < x \leq 5$?

    (b) Which are in the set of values for n when $2 \geq n \geq {}^-3$?

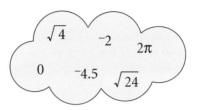

B3 Represent each inequality on a number line.

(a) $2 \leq n \leq 3$ (b) $^-3 < n \leq 1$ (c) $0 \leq n < 6$ (d) $10 > n > 1$

B4 Write inequalities to describe the following diagrams.

(a)
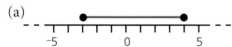

(b)
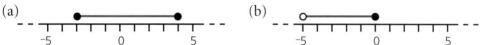

(c)

(d)

B5 If we know that $0 < x < 1$, is each of the following

- always true
- sometimes true
- never true

(a) $x > 0.5$ (b) $^-x < 0$ (c) $2x > 3$ (d) $\frac{1}{x} > 1$

B6 List five different values for n such that $^-1 \leq n < 2$.

B7 List three different values for p such that $8 \leq 2p \leq 10$.

B8 Find two values for x that fit **all** the following inequalities.

$2 > x > {}^-1$ $1 \leq x < 10$ $^-4 \leq x < 1.5$

B9 List all possible integer values of n such that.

(a) $4 \leq n < 9$ (b) $3\frac{1}{3} \leq n < 9$ (c) $^-1 \leq n < {}^-\frac{1}{2}$

(d) $\sqrt{2} < n \leq \pi$ (e) $\frac{1}{4} \leq n \leq \frac{10}{9}$ (f) $1 > n > {}^-1$

(g) $0.4 > n > {}^-2.3$ (h) $3\pi > n > 2\pi$ (i) $^-3.5 \leq x \leq 3.5$

B10 (a) List all the integers n such that $9 \leq n^2 \leq 36$.

    (b) List all the integers n such that $20 < n^2 < 50$.

C In words and in symbols

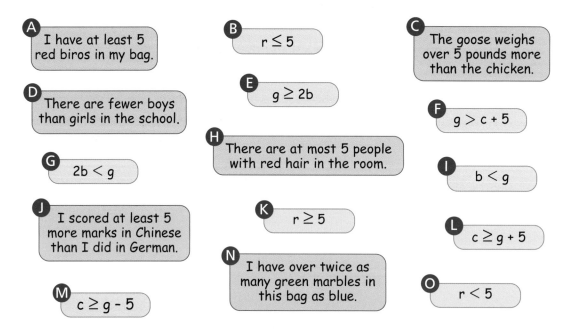

A I have at least 5 red biros in my bag.

B $r \leq 5$

C The goose weighs over 5 pounds more than the chicken.

D There are fewer boys than girls in the school.

E $g \geq 2b$

F $g > c + 5$

G $2b < g$

H There are at most 5 people with red hair in the room.

I $b < g$

J I scored at least 5 more marks in Chinese than I did in German.

K $r \geq 5$

L $c \geq g + 5$

M $c \geq g - 5$

N I have over twice as many green marbles in this bag as blue.

O $r < 5$

C1 Write the following statements using mathematical symbols.

   (a) There are at least 20 pupils in this class. (Use s for the number of pupils.)

   (b) There are over 40 matches in this box. (Use m for the number of matches.)

   (c) There were at most 35 000 spectators at a football match.
      (Use s for the number of spectators.)

   (d) There are more sheep than cows on the farm.
      (Use s for the number of sheep and c for the number of cows.)

   (e) There are more red roses than white and pink roses put together.
      (Use r, w and p for the numbers of red, white and pink roses.)

C2 Match each statement W to Z with one or more of A to J.
In each case, state what any letters stand for.

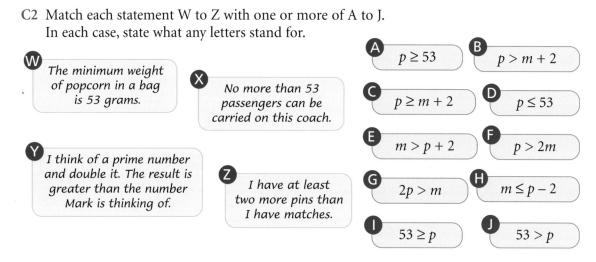

W The minimum weight of popcorn in a bag is 53 grams.

X No more than 53 passengers can be carried on this coach.

Y I think of a prime number and double it. The result is greater than the number Mark is thinking of.

Z I have at least two more pins than I have matches.

A $p \geq 53$

B $p > m + 2$

C $p \geq m + 2$

D $p \leq 53$

E $m > p + 2$

F $p > 2m$

G $2p > m$

H $m \leq p - 2$

I $53 \geq p$

J $53 > p$

C3 Write the following statements using mathematical symbols.
In each case, explain what your letters stand for.

(a) 5 hot-dogs cost more than £3.60.

(b) The number of dogs in the pet shop is less than half the number of cats.

(c) There are at least twice as many ducks as swans on the pond.

(d) 3 bars of chocolate and 4 cans of drink cost the same as 5 burgers.

(e) 3 large bottles of cola are heavier than 7 small bottles of lemonade.

(f) The weight of a chocolate bar is between 190 grams and 210 grams.

(g) You must be over 1.5 metres tall to go on this ride.

(h) There are at least 5 more dogs than cats in the kennels.

(i) The time John took to run 100 metres was at least 12 seconds but under 17 seconds.

C4 Make up statements in words that could match the following inequalities.
In each case, state what the letters stand for.

(a) $g > 5w$

(b) $q \leq g - 7$

(c) $z + p < 20$

(d) $h = 3s + 15$

(e) $7k \geq 140$

(f) $f < 3s$

C5 A printing firm prints leaflets.
It charges a fixed charge of £20 plus £6 per pack of one hundred leaflets.

(a) (i) What is the cost of printing 5 packs of leaflets?

(ii) What is the cost of printing n packs of leaflets?

(b) (i) A youth club has a maximum of £100 to spend on printing some leaflets.
Write down an inequality using n for the number of packs of leaflets.

(ii) What is the largest number of packs the youth club can get printed?

C6

> Children under 2 years go FREE!
> ~~~~~~~~~~~~~~~~~
> Children 12 years or over pay full fare
> ~~~~~~~~~~~~~~~~~
> All other children pay HALF FARE

Write down an inequality to show the ages of children who pay half fare.

C7

**VERY SUPERIOR
CANAL BOATS**

£30 Deposit plus only
£3.50 per hour

John and Susan have £60.

Write down the inequality for the number of hours, t, they can hire a boat for.

What is the longest possible time they can hire a boat for if a boat can only be hired for a whole number of hours?

D Challenges

*D1 Find the highest integer n such that $n^3 \leq 100$.

*D2 What is the lowest integer n such that $n^n > 100\,000$?

*D3 Find the integer n such that $n < \sqrt{200}$ and also $n + 1 > \sqrt{200}$.

*D4 Find t, the smallest possible multiple of 3 for which $t^2 > 200$.

*D5 What is the lowest integer n such that $(\frac{1}{5})^n < \frac{1}{200}$?

*D6 What is the largest prime number p such that $p^2 \leq 500$?

*D7 What is the lowest integer x such that $x^2 < 17$?

*D8 List all the positive integers k for which $2^k < 100$.

*D9 List all the integers n for which $10 \leq n^4 \leq 1000$.

*D10 List all the integers n for which $100 \leq 4^n \leq 10\,000$.

*D11 You have T pence in your pocket made up of only 5p pieces.
If $10 < T \leq 50$, what are the possible values of T?

*D12 You are told that $^-4 \leq m \leq 2$ and that $^-5 \leq n \leq 3$. Find
  (a) the lowest possible value of m^2
  (b) the highest possible value of n^2
  (c) the highest possible value of mn
  (d) the lowest possible value of mn

*D13 Find a fraction r such that $\frac{2}{5} < r < \frac{3}{5}$.

*D14 If a is an integer and $\frac{1}{4} \leq \frac{a}{8} \leq 1\frac{3}{8}$, list all possible values of a.

*D15 What can you say about the smallest possible value of x when $6.1 < x \leq 7.9$?

*D16 The length of a rectangle is a cm and the breadth is b cm.
Write, in terms of a and b, an inequality for each of these statements.
  (a) The area of the rectangle is greater than $30\,\text{cm}^2$ and less than $40\,\text{cm}^2$.
  (b) The perimeter of the rectangle is greater than $50\,\text{cm}$ and less than $70\,\text{cm}$.
  (c) The length of each diagonal of the rectangle is greater than $5\,\text{cm}$ and less than $10\,\text{cm}$.

What progress have you made?

Statement

I can use simple inequalities.

Evidence

1 Decide if each of the following is true or false.

   (a) $7 > {}^-5$ (b) $\sqrt{5} < 2$ (c) $3\pi \geq 9$

2 Represent each inequality on a number line.

   (a) $n \leq 3$ (b) $x > 5$ (c) ${}^-8 \leq y$

3 Write an inequality for

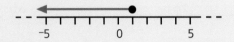

4 Find the largest integer n so that $n^2 \leq 200$.

I can combine simple inequalities.

5 Combine the following pairs of inequalities.

   (a) $4 \leq x$ and $x < 10$

   (b) $n > {}^-5$ and $n < 3$

6 Represent each inequality on a number line.

   (a) $0 < x < 5$ (b) ${}^-5 < n \leq {}^-1$

7 Write an inequality for

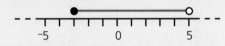

8 List four values for x so that $3 \leq 3x < 10$.

9 If n is an integer, list all possible values of n for $1 < n \leq 2\pi$.

I can write statements in words as inequalities in symbols.

10 Write the following statements using mathematical symbols.
State clearly what any letters stand for.

   (a) There are at least 40 sweets in a bag.

   (b) There are more black sweets than red sweets in a bag.

   (c) The weight of a bag of sweets is at least 100 grams but less than 120 grams.

   (d) There are at least twice as many green sweets as red in a bag.

⟨20⟩ More manipulation

This work will help you

◆ simplify expressions like $n(n + 6) - n(n - 7)$

◆ simplify expressions like $(n + 3)(n - 6)$

◆ use algebra to solve problems and give explanations

A Simplifying: review

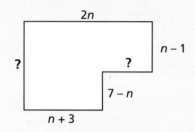

Missing horizontal length $= 2n - (n + 3)$
$= 2n - n - 3$
$= n - 3$

Missing vertical length $= (7 - n) + (n - 1)$
$= 7 - n + n - 1$
$= 6$

Red area $= 10x - 2(x - 3)$
$= 10x - (2x - 6)$
$= 10x - 2x + 6$
$= 8x + 6$

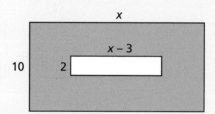

A1 For each expression, multiply out the brackets and collect any like terms.

(a) $5p + 3(p + 1)$ (b) $9x + 2(x - 3)$ (c) $14 + 3(1 - n)$

A2 Write each expression without brackets and collect any like terms.

(a) $4p - (p + 2)$ (b) $20 - (m + 5)$ (c) $12h - (8h - 1)$

(d) $10 - (2h - 3)$ (e) $5x - (1 - x)$ (f) $10 - (7 - 2y)$

A3 Copy and complete each statement.

(a) $\blacksquare - (3 - 2y) = 8y - 3$ (b) $\blacksquare - (2p - 1) = 7 - 2p$

A4 For each expression, multiply out the brackets and collect any like terms.

(a) $^-3(x + 4)$ (b) $^-2(4 - 3a)$ (c) $10h - 7(h + 1)$

(d) $20 - 5(b - 1)$ (e) $8n - 4(3 + 2n)$ (f) $10 - 5(3d - 2)$

A5 Copy and complete each statement.

(a) $10 - 3(2x + \blacksquare) = 7 - 6x$ (b) $6k - 3(1 + \blacksquare) = ^-3$

(c) $\blacksquare - 2(5 - 3y) = 20y - 10$ (d) $\blacksquare - 3(2p + 1) = 23 - 6p$

B Further simplification

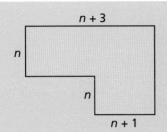

Yellow area $= n(n + 3) + n(n + 1)$
$= n^2 + 3n + n^2 + n$
$= 2n^2 + 4n$

Green area $= 2n(n + 3) - n(n - 1)$
$= (2n^2 + 6n) - (n^2 - n)$
$= 2n^2 + 6n - n^2 + n$
$= n^2 + 7n$

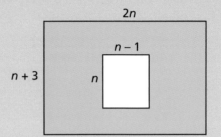

B1 Simplify each expression by collecting any like terms.

(a) $x^2 + 5x + 3x^2$ (b) $6h^2 + h - 2h^2 + 5$ (c) $6m^2 + 2m + 8 - 5m + 1$

(d) $3n^2 + 10 - 4n^2$ (e) $b^2 - 3b + b^2 - b$ (f) $4k^2 + 5 - 5k - 3k^2 - k$

B2 For each expression, multiply out the brackets and collect any like terms.

(a) $4n + n(n - 5)$ (b) $3m(m + 5) - m^2$ (c) $40 + h(5 - h)$

(d) $7b^2 - b(3b + 1)$ (e) $10x - x(x - 7)$ (f) $12g - 4g(3 - g)$

B3 For each expression, multiply out the brackets and collect any like terms.

(a) $n(n + 1) + n(n + 3)$ (b) $b(b - 1) + 5(b - 1)$

(c) $2m(m + 1) + m(m - 5)$ (d) $3y(y - 5) + y(2y - 3)$

B4 For each expression, multiply out the brackets and collect any like terms.

(a) $2d(d + 1) - d(d + 3)$ (b) $n(n + 4) - n(n + 1)$

(c) $x(x + 3) - x(5 - x)$ (d) $p(p - 3) - 5(p - 3)$

B5 Find and simplify expressions for the coloured areas.

(a)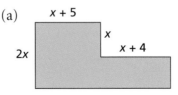

(b)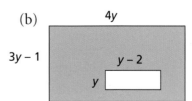

B6 Copy and complete each statement.

(a) $y(y + 3) + \blacksquare(y + 3) = y^2 + 8y + \blacksquare$

(b) $\blacksquare(d - 1) + \blacksquare(d - 1) = d^2 + 4d - \blacksquare$

(c) $\blacksquare(x + 1) - \blacksquare(x + 5) = x^2 - 3x$

C Bracket pairs

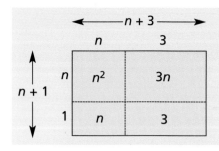

Yellow area $= (n + 1)(n + 3)$
$= n^2 + 3n + n + 3$
$= n^2 + 4n + 3$

$(n + 2)(n + 8)$ ▶▶

	n	8
n	n^2	$8n$
2	$2n$	16

▶▶ $(n + 2)(n + 8)$
$= n^2 + 8n + 2n + 16$
$= n^2 + 10n + 16$

$(n + 2)(n + 8)$
$= n(n + 8) + 2(n + 8)$
$= n^2 + 8n + 2n + 16$
$= n^2 + 10n + 16$

$(n + 2)(n + 8)$ and $n^2 + 10n + 16$
are equivalent expressions
so
$(n + 2)(n + 8) = n^2 + 10n + 16$
is an **identity**.

C1 For each blue rectangle, write an expression for the area
- with brackets – for example, $(n + 2)(n + 8)$
- without brackets – for example, $n^2 + 10n + 16$

(a) $n + 5$ / $n + 2$

(b) $n + 3$ / $n + 3$

(c) $n + 10$ / $n + 1$

C2 Multiply out the brackets from each expression and write the result in its simplest form.
(a) $(x + 2)(x + 6)$
(b) $(x + 12)(x + 1)$
(c) $(x + 4)(x + 1)$
(d) $(x + 2)(x + 2)$
(e) $(x + 6)(x + 3)$
(f) $(x + 2)(x + 9)$

C3 (a) Copy and complete $(x + 5)^2 = (x + 5)(x + 5) =$
(b) Multiply out $(x + 7)^2$ and write the result in its simplest form.

C4 A pupil has tried to multiply out $(x + 4)^2$.

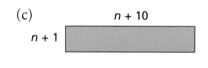

(a) Explain the mistake you think she has made.
(b) Multiply out $(x + 4)^2$ correctly.

C5 Find pairs of expressions from the bubble that multiply to give
(a) $n^2 + 7n + 12$
(b) $n^2 + 6n + 5$
(c) $n^2 + 7n + 6$
(d) $n^2 + 5n + 6$

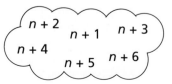

C6 Find expressions for the unknown sides in these rectangles.

(a)

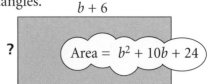

$a + 1$

Area = $a^2 + 8a + 7$

?

(b)

$b + 6$

?

Area = $b^2 + 10b + 24$

C7 (a) Find an expression for the unknown side in this rectangle.

(b) What can you say about this shape?

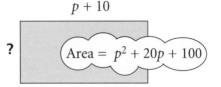

$p + 10$

?

Area = $p^2 + 20p + 100$

C8 Copy and complete these identities.

(a) $(n + 5)(\quad) = n^2 + 8n + 15$

(b) $(\quad)(n + 3) = n^2 + 10n + 21$

(c) $(k + 2)(\quad) = k^2 + \Box k + 22$

(d) $(\quad)(k + 4) = k^2 + 6k + \Box$

C9 Copy and complete these identities.

(a) $(\quad)(\quad) = a^2 + 3a + 2$

(b) $(\quad)(\quad) = b^2 + 11b + 24$

(c) $(\quad)(\quad) = c^2 + 2c + 1$

(d) $(\quad)(\quad) = d^2 + 15d + 26$

C10 (a) Try to complete $n^2 + 2n + 5 = (n + \ldots)(n + \ldots)$ using integers. What do you find? Can you explain this?

(b) Choose your own positive integer values for a and b and try to complete $n^2 + an + b = (n + \ldots)(n + \ldots)$ using integers.

What can you say about a and b when this is possible?

D More brackets

A $(x - 4)(x + 5) = ?$ **B** $(x - 6)(x + 3) = ?$ **C** $(x - 1)(x - 3) = ?$

D1 Multiply out the brackets from each expression and write the result in its simplest form.

(a) $(x - 2)(x + 5)$

(b) $(x + 9)(x - 1)$

(c) $(x - 6)(x + 3)$

(d) $(x + 7)(x - 7)$

(e) $(x - 3)(x - 4)$

(f) $(x - 5)(x - 1)$

(g) $(x + 1)(x - 1)$

(h) $(x - 2)^2$

(i) $(x - 3)^2$

D2 Find pairs of expressions from the bubble that multiply to give

(a) $n^2 + 2n - 3$

(b) $n^2 - n - 2$

(c) $n^2 - 5n + 6$

(d) $n^2 - 9$

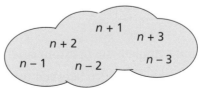

$n + 1$
$n + 2$
$n + 3$
$n - 1$
$n - 2$
$n - 3$

D3 Copy and complete these identities.

(a) $(n + 5)(\quad) = n^2 + 2n - 15$

(b) $(\quad)(m - 6) = m^2 - \Box m + 18$

(c) $(\quad)(\quad) = a^2 - 7a + 10$

(d) $(\quad)(\quad) = b^2 - 4b - 12$

(e) $(\quad)(\quad) = c^2 - 7c - 8$

(f) $(\quad)^2 = d^2 - 10d + 25$

137

D4 Decide which of these statements are identities.

A $(n + 1)^2 = n^2 + 2n + 1$ B $(n - 1)^2 = 9$

C $(n + 2)(n + 3) = n^2 + 6n + 5$ D $(n + 2)(n - 1) = n^2 + n - 2$

E Number patterns and proof

- Copy and complete each pattern.
- What do you notice?
- Do you think it will be true for every possible line?
- Can you prove it's true for every possible line?

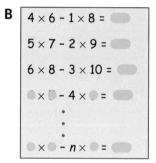

$2 \times 3 - 1 \times 4 = 2$

$3 \times 4 - 2 \times 5 = \bigcirc$

$4 \times 5 - 3 \times 6 = \bigcirc$

$\bigcirc \times \bigcirc - 4 \times \bigcirc = \bigcirc$

$\vdots$

$\bigcirc \times \bigcirc - 100 \times \bigcirc = \bigcirc$

$3 \times 6 - 1 \times 7 = 11$

$4 \times 7 - 2 \times 8 = \bigcirc$

$5 \times 8 - 3 \times 9 = \bigcirc$

$6 \times \bigcirc - \bigcirc \times \bigcirc = \bigcirc$

$\vdots$

$\bigcirc \times \bigcirc - 50 \times \bigcirc = \bigcirc$

E1 For each of these patterns

(a) Copy and complete the first four lines.

(b) Write and simplify the expression for the nth line.

(c) What does this prove about the pattern?

A
$4 \times 4 - 1 \times 7 = \bigcirc$

$5 \times 5 - 2 \times 8 = \bigcirc$

$6 \times 6 - 3 \times 9 = \bigcirc$

$\bigcirc \times \bigcirc - 4 \times \bigcirc = \bigcirc$

$\vdots$

$\bigcirc \times \bigcirc - n \times \bigcirc = \bigcirc$

B
$4 \times 6 - 1 \times 8 = \bigcirc$

$5 \times 7 - 2 \times 9 = \bigcirc$

$6 \times 8 - 3 \times 10 = \bigcirc$

$\bigcirc \times \bigcirc - 4 \times \bigcirc = \bigcirc$

$\vdots$

$\bigcirc \times \bigcirc - n \times \bigcirc = \bigcirc$

E2 (a) Copy and complete the first three lines of this pattern.

(b) Write an expression for the nth line and simplify it.

(c) What does this prove about the pattern?

$6 \times 3 - 8 \times 1 =$

$7 \times 4 - 9 \times 2 =$

$8 \times 5 - 10 \times 3 =$

E3 Make up a pattern of your own where every line gives a result of 6.

E4 (a) Copy and complete the first three lines of this pattern.

(b) Write an expression for the nth line and simplify it.

(c) Explain how this proves that the result for each line will be even.

$7 \times 5 - 1 \times 9 =$

$8 \times 6 - 2 \times 10 =$

$9 \times 7 - 3 \times 11 =$

*E5 (a) Copy and complete the first four lines of this pattern.

(b) Prove that the difference between two consecutive square numbers is always odd.

$2 \times 2 - 1 \times 1 = 3$

$3 \times 3 - 2 \times 2 =$

$4 \times 4 - 3 \times 3 =$

*E6 Prove that the sum of any two consecutive square numbers is always odd.

F Opposite corners

This grid of numbers has ten columns.

A 3 by 3 square outlines some numbers.

Find the 'opposite corner number' like this.

1	2	3	4	5	6	7	8	9	10
11	12	13	⑭	15	⑯	17	18	19	20
21	22	23	24	25	26	27	28	29	30
31	32	33	㉞	35	㊱	37	38	39	40

- Multiply the numbers in opposite corners and find the difference.

  $16 \times 34 - 14 \times 36 = 544 - 504 = \mathbf{40}$

F1 (a) Find the opposite corners number for 3 by 3 squares in different positions on this grid.
What do you think is true about these opposite corners numbers?

  (b) Prove your conclusion.

  (c) Investigate for squares of different size on this grid.

F2 Investigate for grids with different numbers of columns.

G True, iffy, false?

G1 In each statement, n is a positive integer.
Which are
  - always true
  - sometimes true
  - never true

(a) $n^2 + 3n + 1$ is prime

(b) $(n + 1)(n + 2)$ is even

(c) $(n - 1)(n + 2)$ is prime

(d) $(n + 1)(n + 10)$ is odd

(e) $n^2 + n + 1$ is odd

(f) $n^2 + 2n + 1$ is a square number

(g) $(n + 4)(n - 1)$ is a multiple of 5

(h) $(n + 1)(n + 7)$ is a square number

(i) $n^2 + 5n + 6$ is prime

(j) $n^2 + 10n + 24$ is a square number

H Mixed questions

H1 The coloured area here is the difference between a^2 and b^2.
Its two parts can be put together to make a rectangle.

  (a) Find an expression for each side of the rectangle.

  (b) Write down an expression for the area of the rectangle, and hence show that $a^2 - b^2 = (a + b)(a - b)$.

  (c) Multiply out $(a + b)(a - b)$ and simplify your expression to confirm the result in (b).

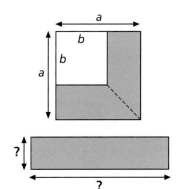

H2 (a) Rewrite the following equation by multiplying out the brackets.

$$(x + 3)(x + 4) = x^2 + 40$$

(b) By subtracting x^2 from both sides of the equation, solve the equation.

H3 Solve these equations.

(a) $(x + 5)(x + 4) = x^2 + x + 52$ (b) $(x + 6)(x - 5) = x^2 - 2$

(c) $(x + 4)(x - 2) = (x + 1)(x + 2)$ (d) $(x + 3)^2 = x^2 - 15$

H4 O is the centre of a circular wheel of radius r cm.
The wheel touches the horizontal ground at P.
PT = 9 cm.
OT crosses the circle at Q and QT = 5 cm.

Use Pythagoras to form an equation for r.
Solve the equation.

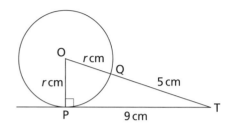

What progress have you made?

Statement

Evidence

I can simplify expressions like $n(n - 2) + 4(n - 2)$.

1 For each expression, multiply out the brackets and collect any like terms.

(a) $n(n - 2) + 4(n - 2)$

(b) $2p(p + 4) - p(p - 3)$

I can multiply out brackets in expressions like $(n + 4)(n + 3)$.

2 Multiply out the brackets from each expression and write the result in its simplest form.

(a) $(n + 5)(n + 4)$ (b) $(x + 8)(x + 1)$

3 Find an expression for the unknown side in this rectangle.

?   Area $= a^2 + 9a + 14$

$a + 7$

I can multiply out brackets in expressions like $(n - 4)(n + 3)$.

4 Multiply out the brackets from each expression and write the result in its simplest form.

(a) $(n + 6)(n - 4)$ (b) $(x - 9)(x - 2)$

I can use algebra to prove general statements.

5 Use algebra to show that the nth line of this pattern will always be equivalent to 5.

$2 \times 6 - 1 \times 7 =$

$3 \times 7 - 2 \times 8 =$

$4 \times 8 - 3 \times 9 =$

$\vdots$

㉑ Quadratic sequences

This work will help you

♦ investigate sequences

♦ form sequences from a quadratic rule

♦ find the rule for a given quadratic sequence

A Sequences

A1 Find the next two terms of each of these sequences.
Can you write down a rule for the next term if you know the previous ones?
(Do not spend too much time on the harder ones.)

(a) 3, 7, 11, 15, …, …

(b) 6, 13, 20, 27, …, …

(c) 4, 5, 7, 10, 14, …, …

(d) 29, 24, 19, 14, …, …

(e) 2, 5, 10, 17, …, …

(f) 6, 12, 24, 48, …, …

(g) 3, 7, 15, 31, …, …

(h) 3, 10, 21, 36, 55, …, …

A2 The formula for the nth term of a particular sequence is $3n + 1$.
So, for example, the 5th term of the sequence is $3 \times 5 + 1 = 16$.

(a) Copy and complete this table of the first eight terms of the sequence.

n	1	2	3	4	5	6	7	8
$3n + 1$	4				16			

(b) By solving the equation $3n + 1 = 76$, work out which term has the value 76.

A3 (a) Write down the first five terms of the sequence whose nth term is $6n - 5$.

(b) For what value of n is the nth term of this sequence 97?

A4 (a) Write down the first five terms of the sequence whose nth term is $n^2 - 1$.

(b) For what value of n is the nth term of this sequence 120?

A5 (a) Write down the first five terms of the sequence whose nth term is $\dfrac{60}{n}$.

(b) For what value of n is the nth term of this sequence   (i) 5   (ii) 0.5

A6 (a) Write down (in their simplest form) the first five terms of the sequence whose nth term is $\dfrac{n-1}{n+1}$.

(b) For what value of n is the nth term of this sequence equal to $\frac{3}{4}$?

B Linear sequences

The formula for the nth term of a **linear sequence** is of the form $an + b$.

Examples are $5n + 2,\ 9n - 5,\ \frac{1}{2}n + 6,\ {}^{-}2n + 7$

B1 (a) The nth term of a sequence is $4n + 3$.
Copy and complete this table to show the first eight terms of the sequence and the differences between each term and the next.

n	1	2	3	4	5	6	7	8
Sequence $4n + 3$	7	11	15	...	...	...	...	...
Differences		4	4	...	...	...	...	...

(b) Make a similar table for the sequence whose nth term is $5n - 1$.

(c) Repeat for the sequence whose nth term is $7n + 4$.

(d) Without making a table, predict what the differences will be for the sequence whose nth term is

(i) $2n + 6$ (ii) $3n - 2$ (iii) $8n + 5$ (iv) $3 + 2n$ (v) $8 - 4n$

B2 (a) Copy and complete this table showing the first eight terms of a linear sequence.

n	1	2	3	4	5	6	7	8
Sequence	8	14	20	26	32	38	44	50
Differences		6	...	...	...	...	...	...

(b) The differences are all 6, showing that the formula for the nth term is of the form $6n + $ some number.

When $n = 1$ the formula gives the first term, which is 8.
Use this to find the number in the formula, and write down the complete formula.

(c) Check your formula by using it to work out some of the other terms of the sequence.

B3 All but one of the sequences below are linear.
Use the differences method to find the formula for the nth term of each linear sequence.
Try to find the formula for the nth term of the non-linear sequence.

(a)

n	1	2	3	4
Sequence	5	9	13	17

(b)

n	1	2	3	4
Sequence	3	9	15	21

(c)

n	1	2	3	4
Sequence	5	5.5	6	6.5

(d)

n	1	2	3	4
Sequence	4	1	${}^{-}2$	${}^{-}5$

(e)

n	1	2	3	4
Sequence	${}^{-}1$	${}^{-}5$	${}^{-}9$	${}^{-}13$

(f)

n	1	2	3	4
Sequence	2	5	10	17

C Second differences

Sequence	2		6		12		20		30		42		56		72
First differences		4		6		8		...		...		...		...		
Second differences			2		2		...		...		...		...		...	

C1 Use rows of differences to investigate the following sequences.
Find the next two terms which follow the patterns you have noticed.

(a) 4, 6, 9, 13, 18, ..., ...

(b) 4, 10, 18, 28, 40, 54, ..., ...

(c) 0, 7, 18, 33, 52, ..., ...

(d) 3.4, 4.7, 6.2, 7.9, ..., ...

C2 In each of these sequences the second differences are constant.
Copy and complete the rows to find the first six terms of each sequence.

(a)

n	1		2		3		4		5		6
Sequence	...		...		...		20		...		...
First differences		4		6		8		...		...	
Second differences			...		...		...		...		

(b)

n	1		2		3		4		5		6
Sequence	...		...		...		30		...		...
First differences		3		7		...		...		...	
Second differences			...		...		...		...		

(c)

n	1		2		3		4		5		6
Sequence	...		...		...		27		...		...
First differences		...		...		3		...		1	
Second differences			...		...		...		...		

(d)

n	1		2		3		4		5		6
Sequence	...		...		...		...		...		50
First differences		12		...		...		...		10	
Second differences			...		...		...		...		

C3 These sequences do not have constant second differences.
But you can still use differences to help find the next terms.
Find the next two terms of each sequence.

(a) 3, 10, 29, 66, 127, 218, ..., ...   (b) 0, 1, 4, 10, 20, 35, ..., ...

D Investigating quadratic sequences

In a **quadratic sequence**, the nth term is given by a formula of the form $an^2 + bn + c$.
Examples are n^2, $3n^2 + 6$, $5n^2 - 2n + 4$, $-2n^2 + 8n$.

- Copy and complete this table for the sequence whose nth term is n^2.

n	1	2	3	4	5	6	7
Sequence	1	4	9	16	25	36	49
First differences		3	5	...	...	...	...
Second differences			...	...	...	...	...

- Copy and complete this table for the sequence whose nth term is $3n^2$.

n	1	2	3	4	5	6	7
Sequence	3	12	27	48	...	...	...
First differences		9	15	...	...	...	...
Second differences			...	...	...	...	...

- Investigate the row of second differences for other sequences whose nth term is of the form an^2.
  What is the connection between the value of a and the second differences?

- Copy and complete this table for the sequence whose nth term is $3n^2 + 7$.

n	1	2	3	4	5	6	7
Sequence	10	19	34	55	...	...	...
First differences		9	15	21	...	...	...
Second differences			...	...	...	...	...

- Does adding 7 to the formula $3n^2$ have any effect on the second differences?
- Investigate the row of second differences in other pairs of sequences of the form an^2 and $an^2 + c$.
  What do you find?
- Investigate pairs of sequences of the form an^2 and $an^2 + bn$
  (for example, $4n^2$ and $4n^2 + 3n$).
  What effect does the term '+ bn' have on the second differences?
- For any sequence of the form $an^2 + bn + c$, what do the second differences tell you about a, b or c?

E Finding the formula for a quadratic sequence

The nth term of a quadratic sequence can be written as $an^2 + bn + c$.
The second differences are constant and are equal to $2a$.

We can use this to help find the formula for a quadratic sequence.

E1 (a) Copy and complete the following table for the sequence 4, 15, 32, 55, ...

n	1	2	3	4	5	6	7
Sequence	4	15	32	55	84	119	160
First differences		11	17	...	...	...	...
Second differences			6	...	...	...	...

(b) Check that all the second differences are 6.
This suggests that the formula for the nth term is $3n^2$ + something.

To find the something, we form a new sequence by subtracting $3n^2$ from each term of the original sequence. The table below shows this.

n	1	2	3	4	5	6	7
Sequence	4	15	32	55	84	119	160
$3n^2$	3	12	27	48	75	108	147
Sequence – $3n^2$	1	3	5	7	9	11	13

Find the first differences for this new sequence 1, 3, 5, 7, ...
Hence find a formula for the nth term of this new sequence.

(c) Use the fact that you subtracted $3n^2$ from each term of the original sequence to write down a formula for the nth term of the original sequence.

E2 (a) These are the first seven terms of a quadratic sequence.
Copy and complete the table.

n	1	2	3	4	5	6	7
Sequence	7	17	31	49	71	97	127
First differences		10	14	...	...	...	...
Second differences			...	...	...	...	...

(b) Check that all the numbers in your second row of differences are constant.
If the nth term of the sequence is an^2 + something,
what does this tell you about a?

(c) Form a new sequence in a copy of the table below by subtracting an^2 from each term. Use the value of a you found in part (b).

n	1	2	3	4	5	6	7
Sequence	7	17	31	49	71	97	127
an^2	...	...	...	...	...	...	...
Sequence – an^2	...	...	...	...	...	...	...

(d) Find a formula for the nth term of the new sequence.
Use it to write down a formula for the nth term of the original sequence.

(e) Use your formula to work out the 8th and 9th terms of the original sequence.
Check your answers by continuing the difference table
in part (a) for two more terms to the right.

E3 Use the method above to find the nth term of each of these sequences.
In each case, work out the next two terms using your formula
and check using your difference table.

(a) 7, 22, 45, 76, 115, 162, 217, ...

(b) 6, 11, 20, 33, 50, 71, 96, ...

(c) 0, 17, 42, 75, 116, 165, ...

(d) 0, 3, 10, 21, 36, 55, 78, ...

(e) ⁻3, 7, 23, 45, 73, ...

(f) 1, $3\frac{1}{4}$, 7, $12\frac{1}{4}$, 19, ...

*E4 Here are the first few terms of some sequences which are **not** quadratic.
Can you spot what the nth term might be for each one?

(a) 1, 8, 27, 64, 125, 216, ...

(b) 3, 10, 29, 66, 127, 218, ...

(c) 2, 4, 8, 16, 32, 64, 128, ...

(d) 10, 100, 1000, 10 000, 100 000, ...

(e) 11, 102, 1003, 10 004, 100 005, ...

(f) 11, 104, 1009, 10 016, 100 025, ...

(g) $\frac{1}{1}, \frac{1}{2}, \frac{1}{3}, \frac{1}{4}, \frac{1}{5}, \frac{1}{6}, \ldots$

(h) $\frac{1}{2}, \frac{1}{3}, \frac{1}{4}, \frac{1}{5}, \frac{1}{6}, \frac{1}{7}, \ldots$

(i) $\frac{1}{2}, \frac{2}{3}, \frac{3}{4}, \frac{4}{5}, \frac{5}{6}, \frac{6}{7}, \frac{7}{8}, \ldots$

(j) $\frac{1}{1}, \frac{3}{4}, \frac{5}{7}, \frac{7}{10}, \frac{9}{13}, \frac{11}{16}, \ldots$

What progress have you made?

Statement	Evidence
I can use first and second differences to investigate a sequence.	1 Find the first two rows of differences and the next two terms for the following sequence. 2, 2, 4, 8, 14, ..., ...
I can form sequences from a quadratic rule.	2 Find the first five terms of the sequence whose nth term is $3n^2 - 4n + 1$
I can find a rule for the nth term of a quadratic sequence.	3 For the following quadratic sequence, find the first two rows of differences and hence find a formula for the nth term. 8, 26, 54, 92, 140, 198, ...

Review 3

1. Rewrite each of these formulas in the form $x = \dots$

   (a) $y = 3x + 5$ (b) $y = \frac{x-4}{3}$ (c) $y = \frac{x}{5} + 7$

2. This design is to be enlarged so that the length marked p becomes 18 cm.

   (a) What is the scale factor of the enlargement?

   (b) What does the length q become in the enlargement?

   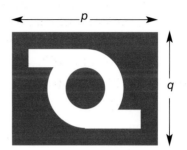

3. The equation of the graph P is
   $$y = \tfrac{1}{2}x^2 - 3$$

   (a) Use this graph to estimate the solutions of the equations

   (i) $\tfrac{1}{2}x^2 - 3 = 2$

   (ii) $\tfrac{1}{2}x^2 = 3$

   (b) What is the equation of graph Q?

   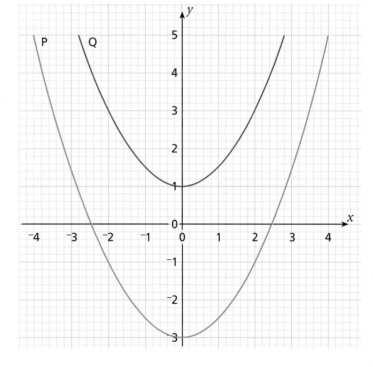

4. Write down the inequality shown by each of these diagrams.

   (a)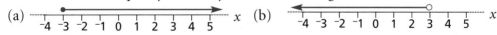

   (b)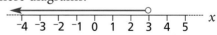

   (c)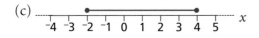

   (d)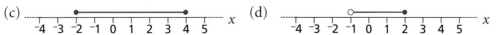

5. Multiply out the brackets and simplify each of these.

   (a) $3(x - 5) - 2(6 - 5x)$ (b) $(x + 4)(x + 7)$ (c) $(x - 3)(x + 9)$

6 (a) Write down the formula for the nth term of this sequence.

n	1	2	3	4	5
Sequence	1	4	9	16	25

(b) Hence write down the formula for the nth term of each of these sequences.

(i)

n	1	2	3	4	5
Sequence	31	34	39	46	55

(ii)

n	1	2	3	4	5
Sequence	$\frac{1}{2}$	2	$4\frac{1}{2}$	8	$12\frac{1}{2}$

(iii)

n	1	2	3	4	5
Sequence	1	$2\frac{1}{2}$	5	$8\frac{1}{2}$	13

(iv)

n	1	2	3	4	5
Sequence	0	1	4	9	16

7 This shape consists of a square of side s cm with a semicircle of diameter s cm on each of two sides.

The perimeter of the shape is P cm and the area is A cm².

(a) Explain why $P = (\pi + 2)s$.

(b) Write a formula for s in terms of P.

(c) Explain why $A = (\frac{\pi}{4} + 1)s^2$.

(d) If $A = 45$, find s correct to 1 d.p.

8 (a) Calculate the ratio $\dfrac{\text{height}}{\text{width}}$ for this design.

(b) The design is to be enlarged so that the width becomes 19.5 cm.
Calculate the enlarged height.

height 2.4 cm

width 6.0 cm

9 List all the integers n that satisfy each of these inequalities.

(a) $^-3 < n < 4$　　　　**(b)** $^-2.5 \le n \le 1.5$　　　　**(c)** $0 \le n^2 \le 15$

10 (a) Write the next two lines of this pattern.

(b) Write the nth line of the pattern.

(c) Prove that the result of every line of the pattern is 12.

$4 \times 5 - 1 \times 8 = ...$
$5 \times 6 - 2 \times 9 = ...$
$6 \times 7 - 3 \times 10 = ...$
$7 \times 8 - 4 \times 11 = ...$

$(...) \times (...) - n \times (...) = ...$

11 Use a calculator to work these out. Give answers to three significant figures.

(a) $\dfrac{9.45 + 2.37}{8.03 - 1.55}$　　　　**(b)** $\dfrac{11.3}{0.47 \times 8.12}$　　　　**(c)** $25.4 - \dfrac{45.2}{18.6 - \sqrt{23.2}}$

12 Calculate the length marked L in this trapezium.

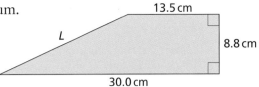

13 Karina is carrying out an experiment with an electric circuit.
   She connects different components across the terminals of a battery.
   She measures the current, I amp, and the voltage, V volt, each time.

   Here is a graph of her results.

   Find the equation linking V and I.

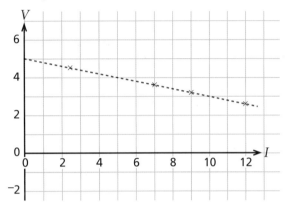

14 This stem-and-leaf table shows the marks scored by some
   students in an examination.

   (a) How many students took the examination?

   (b) What was the range of the marks?

   (c) What was the median mark?

0	7 8 9
1	0 2 4 5 6 6 6 9
2	1 2 2 4 5 6 7 7 8
3	2 4 4 5 7 7 7 8 9 9 9
4	0 0 2 3 4 4 5 5 6

   Later it is discovered that the student whose mark was recorded
   as 10 should have scored 20.

   (d) What is the effect on the median of changing the mark of 10 to 20?

   (e) By how much does the mean mark increase as a result of the change?

15 Copy this diagram on to squared paper.

   (a) Triangle A is rotated 180° about (4, 0).
       The result is rotated 180° about (8, ⁻1).

       Draw the final image and label it B.

   (b) What single transformation maps A to B?

   (c) Triangle A is rotated 90° anticlockwise
       about (4, 4).
       The result is rotated 90° clockwise
       about (6, 1).

       Draw the final image and label it C.

   (d) What single transformation maps C back to A?

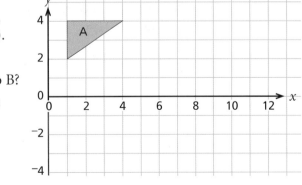

16 (a) What is the total of all the interior angles of a heptagon (7 sides)?

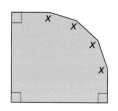

(b) The angles marked x in this heptagon are equal. Calculate x.

17 Solve these equations.

(a) $3(7 - x) + 4(x - 3) = 6$

(b) $2(x - 5) - 3(x - 4) = 7$

(c) $4(2x - 3) = 3(6 - 4x)$

(d) $(x + 6)(x - 3) = x^2 - x$

18 Carol has two boxes of chocolates.
The first box contains 7 dark chocolates and 3 light chocolates.
The second box contains 6 dark chocolates and 3 light chocolates.

Carol takes a chocolate at random from each box.
Find, as a fraction, the probability that

(a) both chocolates are dark

(b) both chocolates are light

(c) one chocolate is light and the other dark

19 Given that $P = 2(a + b)$, express b in terms of P and a.

*20 The fish tank shown here is partially filled with water.
It is tipped up until the water is just about to flow over an edge.
Calculate the length marked x.

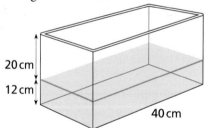

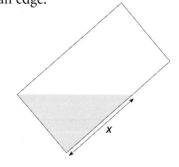

*21 You have 2p and 3p stamps.
You stick them in a single row.

For example, this row 2p 3p 2p 2p is recorded as 2, 3, 2, 2

The table shows all the different ways of making up totals of 2p, 3p, 4p, …

Investigate the sequence in the third column of the table.

Try to explain why it occurs.

Total	Arrangements					Number
2p	2					1
3p	3					1
4p	2,2					1
5p	2,3	3,2				2
6p	2,2,2	3,3				2
7p	2,2,3	2,3,2	3,2,2			3
8p	2,2,2,2	2,3,3	3,2,3	3,3,2		4
9p	2,2,2,3	2,2,3,2	2,3,2,2	3,2,2,2	3,3,3	5

Index